Meteorites, Ice, and Antarctica

Bill Cassidy led meteorite recovery expeditions in the Antarctic for 15 years. His searches resulted in the collection of thousands of meteorite specimens from the ice. This fascinating story is a first hand account of his field experiences on the US Antarctic Search for Meteorites Project, which he carried out as part of an international team of scientists. Cassidy describes this hugely successful field program in Antarctica and its influence on our understanding of the moon, Mars and the asteroid belt. He describes the hardships and dangers of fieldwork in a hostile environment, as well as the appreciation he developed for the beauty of the place. In the final chapters he speculates on the results of the trips and the future research to which they might lead.

BILL CASSIDY was the founder of the US Antarctic Search for Meteorites project (ANSMET). He received the Antarctic Service Medal of the United States in 1979, in recognition of his successful field work on the continent. His name is found attached to a mineral (cassidyite), on the map of Antarctica (Cassidy Glacier) and in the Catalogue of Asteroids (3382 Cassidy). He is currently Emeritus Professor of Geology and Planetary Science at the University of Pittsburgh.

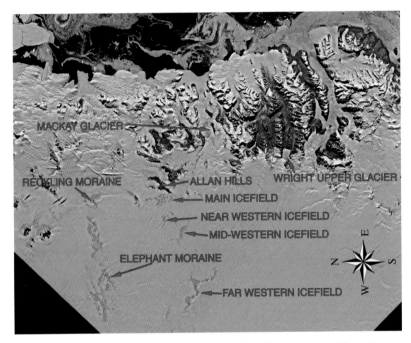

Frontispiece: The illustration shows a digitally enhanced, false-color mosaic of satellite photos of the Allan Hills – Elephant Moraine area. Blue areas are patches of exposed ice. Notice that the Allan Hills Main Icefield stands away from the roughly Y-shaped Allan Hills exposure, due to the presence of a low-lying structural barrier (a subice ridge). Ice flows over this barrier toward Allan Hills. Elephant Moraine is also indicated. The regional linear patches of blue ice, in one of which are found Elephant Moraine and Reckling Moraine, mark the presence of a subice ridge. Ice is spilling over this ridge on its journey northward. The irregular dark area at the top of the photo is open water of the Ross Sea, which is completely frozen during most of the year. Contorted patterns in the water are aggregates of floating ice chunks whose trends reflect eddy currents. Brownish patches in the upper right quadrant are Dry Valleys. (Courtesy of Baerbel Luchitta, USGS Image Processing Facility, Flagstaff, Arizona, USA)

Meteorites, Ice, and Antarctica

WILLIAM A. CASSIDY

University of Pittsburgh

CAMBRIDGE
UNIVERSITY PRESS

PUBLISHED BY THE PRESS SYNDICATE OF THE UNIVERSITY OF CAMBRIDGE
The Pitt Building, Trumpington Street, Cambridge, United Kingdom

CAMBRIDGE UNIVERSITY PRESS
The Edinburgh Building, Cambridge CB2 2RU, UK
40 West 20th Street, New York, NY 10011-4211, USA
477 Williamstown Road, Port Melbourne, VIC 3207, Australia
Ruiz de Alarcón 13, 28014 Madrid, Spain
Dock House, The Waterfront, Cape Town 8001, South Africa

http://www.cambridge.org

First published 2003

Printed in the United Kingdom at the University Press, Cambridge

Typeface Trump Mediaeval 9.5/15 pt   *System* LATEX 2_ε [TB]

A catalogue record for this book is available from the British Library

Library of Congress Cataloging in Publication data

Cassidy, W. A. (William A.)
Meteorites, Ice, and Antarctica / William Cassidy.
    p.   cm. – (Studies in polar research)
Includes bibliographical references and index.
ISBN 0-521-25872-3
1. U.S. Antarctic Search for Meteorites Project.   2. Meteorites – Antarctica.
I. Title.   II. Series.
QB755.5.A6 C37   2003
5235′1′ 09989–dc21   2002073932

ISBN 0 521 25872 3 hardback

I dedicate this book to my wife, Bev, who ran our home, and our family, for fifteen field seasons while I was in Antarctica, and never once complained.

Contents

9   Evaluating the collection – and speculating on
    its significance   227

10 Meteorite stranding surfaces and the ice sheet   274

11 The future: what is, is, but what could be,
    might not   320

Appendices

A   The US–Japan agreement   335

B   ANSMET field participants, 1976–1994   337

    Index of people   342

    Index of Antarctic geographic names   344

    Subject index   346

Foreword

This wonderful tale of physical and intellectual adventure details the development of the ANSMET (Antarctic Search for Meteorites) program of meteorite collection in Antarctica and its importance for planetary science. Starting from the chance discovery by Japanese glaciologists of several different *types* of meteorites in a limited field area of Antarctica, Cassidy describes the flash of insight that led to his conviction that Antarctica must be a place where many meteorites could be found. His basic idea was that it was wildly improbable to find different meteorites in a limited area *unless there was a concentration mechanism at work*. The subsequent discovery of several hundred meteorite samples by another Japanese team proved the point.

Alas, insights are not always easily shared. The initial rejection of his proposal to test his idea serves as a most useful lesson to young scientists everywhere – don't be discouraged by initial rejection of your new ideas, persist!

Initially undertaken as a joint Japanese–American effort, the national programs eventually diverged. The work directed by Cassidy matured into the highly successful ANSMET program that has become an integral part of the NSF's (National Science Foundation) polar research program.

I had the good fortune to participate in two ANSMET field seasons and believe that ANSMET is organized in just the right way. It need not have been thus. I suspect that most of us faced with the problem of collecting meteorites in the hostile Antarctic environment would have opted to send in teams of vigorous young male adventurers. And one would have been tempted to use the specimens so collected for one's personal research. But Cassidy had the wisdom to do things differently. The ANSMET field teams consist of a mixture of young and old, professors and students, male and female, Americans and citizens of other countries, with a sprinkling (mostly John Schutt) of experienced field people termed "crevasse experts". They share a common love for, and knowledge of, the scientific study of meteorites. The inclusion of lab scientists in the field teams has led to a much better understanding of the nature of the samples – it is impossible to speak of "pristine" samples when one has seen a black meteorite sitting in a puddle of melt water!

The meteorites are initially handled at NASA's Johnson Spacecraft Center in Houston, and scientists from all countries are invited to request samples. As with the lunar samples before them, the meteorites are considered as the heritage of the human race as a whole. This is as it should be.

The book shows why meteorites are scientifically interesting and the "intellectually curious general reader" addressed by Cassidy will learn much. A foreword is no place to delve into scientific particulars. Suffice to say that almost everything we *know* (as opposed to hypothesize) about the formation and early history of the Solar System is derived from studies of meteorites.

Most, but not all, meteorites are fragments of asteroids. Two important exceptions are those (rare) meteorites that come from the Moon and from the planet Mars. A major part of the NASA Planetary Science program is the continued exploration of Mars with the goal of one day returning samples of the planet to earth. The total cost will run into many billions of dollars. The continued collection of Martian meteorites from Antarctica, at a tiny fraction of the cost of a sample return mission, is clearly warranted. Cassidy also makes a convincing case of continuing the search for new lunar meteorites.

Museum collections have now been greatly surpassed by the thousands of Antarctic finds. A natural question is whether we really need more meteorites. Cassidy shows why the answer is a resounding yes! As luck would have it, the rate of return of interesting specimens just about matches the rate at which they can be properly studied.

There is thus every reason to continue the existing collection effort at about the same level.

Like most meteoriticists, Cassidy emphasizes the planetary insights gleaned from meteorites. He shows explicitly how the sampling of asteroidal fragments permits the study of the melting and differentiation of small planets leading to a better understanding of the processes that operated on the early earth.

Although not discussed by Cassidy, the reader might be interested to learn that meteorites also provide unique information about the larger universe beyond the planets. Relatively recently, researchers have shown that meteorites contain small grains of interstellar dust that formed around different stars at different times prior to the formation of our sun. The detailed study of these grains, some of which formed in the atmospheres of dying stars similar to our own, and others in supernova explosions, provide new insights into stellar evolution and the processes of element formation. Meteorites also provide unique information about the nature and history of galactic cosmic rays.

Cassidy's discussion of the meteorite concentration mechanism and its possible implications for future studies of past and present Antarctic ice movements is both original and important. In collaboration with the late glaciologist, Ian Whillans, he developed a basic model for "meteorite stranding surfaces." These are envisioned as backwaters of ice flows around natural barriers where wind ablation (wind is a near constant presence in Antarctica) serves to build up the surface concentration of meteorites originally trapped in the volume of the incoming ice. He surmises that measurements of the distribution of terrestrial ages of meteorites on different stranding surfaces, coupled with careful glaciological measurements of current ice flow patterns and sub-surface topography, could give new information on the history of the ice flows. He also signals the potential importance of dust bands in the ice for providing "horizontal ice cores" which, if they could be properly dated, would add to our overall understanding. His ideas deserve to be further exploited.

The book treats grandiose phenomena such as the nature of the Antarctic ice sheet and the march of the ice from the polar plateau to the sea. But it is also a highly personal and intimate account. The reader will see clearly the thought patterns and passions that characterize the natural scientist.

I also trust that the reader will understand why other ANSMET veterans and I find Cassidy to be such a splendid expedition companion. His wonderful sense of humor breaks out repeatedly (and mostly unexpectedly) throughout the narrative. I cite just one example. In trying to understand why the meteorite concentrations were not discovered earlier he realizes the dog teams do very poorly on ice fields and such places were thus avoided. This leads him to speculate on equipping dogs with crampons – a thought quickly dismissed as he imagines the consequences of a crampon-equipped dog scratching its ear! I invite the reader to find and enjoy the many other examples sprinkled throughout the text.

Robert M. Walker
McDonnel Professor of Physics
Washington University
January 2003

Acknowledgments

I hope, and intend, that this book will appeal to the intellectually curious general reader, as well as those who do research on meteorites and field work in Antarctica. In seeking to write such a book, I have prevailed upon the good natures of a number of friends and colleagues to read early drafts, criticize, and suggest. The following persons have done much to influence the final form of the book. I thank them all, very sincerely.

Bev Cassidy, for reading several chapters and making suggestions.

John Schutt, for reading several chapters for accuracy and detail.

Bob Fudali, for reading the entire typescript for style and content.

Mike Zolensky, for suggestions on Chapter 6.

Hap McSween, for critical reading and suggestions on Chapter 6.

Randy Korotev, for critical reading and suggestions on Chapter 7.

Bruce Hapke, for periodic consultations.

Leon Gleser, for critical reading of Chapter 9.

Lou Rancitelli, for critical reading and style suggestions on Chapter 9.

Kunihiko Nishiizumi, for age determinations, before publication.

Parts I and II of this book were reviewed by Roger Hewins and Part III was reviewed by Phil Bland. These were very constructive reviews, containing excellent suggestions. I followed many suggestions but declined others, for one reason or another. If the book is less than it could be because I have not accepted all these suggestions I accept full responsibility.

Introduction

The Yamato Mountain Range wraps the ice sheet around its shoulders like an old man with a shawl. Ice coming from high off the ice plateau of East Antarctica, arriving from as far away as a subice ridge 600 km to the south, finds this mountain range is the first barrier to its flow. The ice has piled its substance up against the mountains in a titanic contest that pits billions of tons of advancing ice against immovable rock, whose roots extend at least to a depth of 30 km. The ice is moving because billions of tons of ice are behind it, pushing it off the continent and into the sea. Ultimately it yields, diverging to flow around the mountains. On the upstream side the rocks have been almost completely overwhelmed – only pink granite peaks protrude above the ice, which spills down between and around them in tremendous frozen streams and eddies, lobes, and deeply crevassed icefalls. The change in elevation of some 1100 m between the high plateau upstream of the mountains and the lower ice flowing away from the downstream slopes creates a spectacular view of this giant downward step in the ice surface. Almost constant howling winds from the interior blow streamers of ice crystals off the mountain peaks and "snow snakes" dance down the slopes in sinuous trains, as if somehow connected to each other. The scale of the scene is such that people become mere specks in an awesome, frigid emptiness.

In 1969, a group of Japanese glaciologists were specks in this scene. With all their supplies and equipment, they had traveled inland 400 km from Syowa Base, on the coast, to reach the Yamato Mountains (called the Queen Fabiola Mts. on most maps) and carry out measurements on the velocity of ice flow, rate of ablation and ice crystallography. Their safety depended on the reliable operation of two tracked vehicles in which they ate, slept and waited out the storms. These scientists were physically hardy and highly motivated. Because the Japanese supply ship could reach Syowa base only in the middle of summer, when parties had already left for the field, they had already wintered over at Syowa Base and would spend another winter there before being able to return to their families, just so they could spend the four months of antarctic summer at this desolate place, gathering fundamental data along the margin of a continental ice sheet. One of them, Renji Naruse, picked up a lone rock that was lying on the vast bare ice surface and recognized it as a meteorite.

In the preceding 200 years only about 2000 different meteorites had been recovered over the entire land surface of the earth, and finding a meteorite by chance must be counted as extremely improbable. It's lucky, therefore, that this initial discovery at the Yamato Mountains was made by a glaciologist, who would not be expected to have a quantitative understanding of exactly how rare meteorites really are, and of what a lucky find this should have been; Naruse and his companions proceeded to search for more. By day's end they had found eight more specimens in a 5×10 km area of ice – a tiny, tiny fraction of the earth's land surface.

Until that time, such a concentration always represented a meteorite that had broken apart while falling through the earth's atmosphere, scattering its fragments over a small area called a strewnfield. In such a case, all the fragments are identifiable as being of the same type. In this instance, however, all nine meteorites were identifiably different, and so were from different falls. A meteoriticist would strike his forehead with the palm of his hand, in disbelief.

Naruse and his companions undoubtedly were pleased with this unexpected addition to their field studies but there is no record that they immediately attached great significance to the find. They bundled up the specimens carefully, for return to Japan, and then resumed the ice studies that had drawn them to this spot. The ice at the Yamato Mountains, however, was destined for great fame, not for its glaciology but for the thousands of meteorites that would later be found on its surface. One might say that the Yamato Mountains icefields were *infested* with meteorites.

This book is about what some of us did about that discovery, how we did it, what we thought while we were doing it, and what the effects have been on planetary research.

Part I Setting the Stage

Antarctica is the best place in the world to find meteorites, but it is also a singular place in many other ways. In Part I, while I outline the manner in which the Antarctic Search for Meteorites (ANSMET) project came into being, I also describe our field experiences as untested beginners, discovering the hardships and dangers of this special place in the world, as well as our slowly growing awareness and appreciation of its alien beauty. Antarctica is a *presence* in any scientific research conducted there, imposing its own rules upon what can and cannot be done, how things can be done, and what the cost is for doing those things. At the same time, it rewards the dedicated field person, not only in yielding scientific results not available anywhere else in the world, but with a headful of wonderful memories, startling in their clarity, of snow plumes swept horizontally off rocky peaks like chimney smoke in a strong wind; of poking a hole through a snowbridge and marveling at the clusters of platy six-sided ice crystals that have grown in the special environment of a crevasse below the fragile protection of a few centimeters of snow; of emerging from one's tent after a six-day storm to find the delicate snow structures randomly sculpted by a wind which, while it was churning furiously through camp, seemed to have no shred of decency about it, much less any hint of an artistic impulse; of returning late one evening after a 12-hour traverse to a campsite occupied earlier in the season, when the sun makes a low angle to the horizon and we camp beneath a tremendous tidal wave of ice with its downsun side in shadow and displaying every imaginable shade of blue, and, having been there before, learning again the pleasant feeling of having come home.

I Antarctica and the National Science Foundation

Antarctica occupies about 9% of the earth's total land surface. For this to be true, of course, you must accept snow and ice as "land surface," because this is what mainly constitutes that part of the continent that lies above sea level. Think of the antarctic continent as a vast convex lens of ice with a thin veneer of snow. In contrast to the region around the north pole, which is just floating ice at the surface of the ocean, the antarctic ice lens rests on solid rock. In most places the ice is so thick, and weighs so much, that it has depressed the underlying rock to about sea level. If the ice melted completely, the surface of the continent would rebound over a long period of time until its average elevation would be higher than any other continent. As it is, the ice surface itself gives Antarctica a higher average elevation than any other continent.

It is only in a very few places, where mountains defy the ice cover, that we can directly sample the underlying rocks. Most of these places are near the coast, where the ice sheet thins. At the center of the continent the elevation is about 4000 m. At the south geographic pole, which is not at the center of the continent, the elevation is 3000 m.

This ice ocean is both vast and deep. Except near the coast, total precipitation averages less than 15 cm of water-equivalent per year, so Antarctica is by definition a desert. It has accumulated such a great thickness of ice by virtue of the fact that whatever snow does fall, doesn't melt. Antarctic ice comprises about 80% of all the fresh water on the earth's surface. This great mass of ice is not contained at its margins, so as it presses downward it ponderously moves outward, creeping away from its central heights toward the edges, thinning and losing altitude as it spreads out, but partly replenished along its way by sparse precipitation.

We have marked the southernmost point on earth with a pole surmounted by a silvered sphere, of the type sometimes seen on well kept lawns or in formal gardens. But ice is moving past the geographic south pole at a rate of 10 m per year, so every few years we must get the pole and bring it back to its proper location. The problem is less tractable for South Pole Station itself. It slowly drifts away with the ice and at the same time sinks ever deeper as the yearly snowfalls impose their will. As a result, we have a string of several buried former South Pole Stations marking the particular flow line that passes through the south pole. They are accessible for a while, but as they go deeper below the surface they are ultimately crushed flat, or invaded and filled by ice.

Field conditions in Antarctica are extreme; more so the closer one approaches the south pole. The areas where we work are typically at 2000 m elevation. In these areas and at the times of year during which we are in the field we expect temperatures ranging between -10 and $-25\,°C$. In still air, with proper clothing and a high-calorie diet, these temperatures are quite tolerable. In moving air they are less so.

We are in Antarctica during the relatively more balmy months of the austral summer: November, December, and January. This is also a time of continuous daylight: suppose when you emerge from your tent in the morning, the sun is shining directly on the entrance. It will be at an elevation in the sky that I would read as around 10 a.m., if I were home in Pennsylvania. During the following 24 hours, due to the rotation of the earth, the sun will appear to make a complete circle of the tent, but will always give the impression that the time of day is around 10 a.m. Actually, at "night" it will appear to be around 9 a.m., changing its angle of elevation a little because we are not exactly at the south pole. But it never sets during the summer season. Knowing this does not mean that we immediately adjust to this new set of conditions. Many times we leave our camp when the atmosphere is hazy, and I find myself thinking, "Well, this fog will burn off as soon as the sun comes up." And the sun has been up for two months!

In the past, territorial claims have been made in Antarctica by Argentina, Australia, Chile, France, New Zealand, Norway and the United Kingdom. Because of sometimes overlapping claims, about 110% of Antarctica was divided up, in pie-shaped areas that converged to points at the south pole. The exception to this was Norway, whose claim stopped at 85° S and looked like a piece of pie that someone had begun to eat. Of the seven countries claiming territory, only Norway stopped short of the south pole, and she seemingly had more right than anyone else to claim it because the Norwegian explorer Roald Amundsen had been the first to reach the south pole.

In an effort to reduce tensions over the expressions of nationalism represented by territorial claims, the claiming nations were persuaded to set aside their aspirations temporarily and, with six other nations, to sign an international accord: the Antarctic Treaty. This treaty has by now been acceded to by 45 nations, and 27 of these are conducting active research programs there. The treaty provides for unhindered access to any part of Antarctica by any signatory nation for scientific purposes. The United States (US) is a signatory nation but makes no territorial claim. We have a large and continuing scientific effort in Antarctica that is supervised by the National Science Foundation (NSF).

MCMURDO STATION

The US has permanent year-round research bases on Ross Island (*McMurdo Station*), at the South Pole (*Amundsen–Scott South Pole Station*), and on the Antarctic Peninsula (*Palmer Station*) (see map, Fig. 1.1). By far the largest of these is McMurdo. At 77° 30′ S, it is admirably sited for scientific work, being as far south as is practical for late-summer access by small ocean-going vessels aided by an ice-breaker, so that yearly resupply missions can be relied upon. It is at the land–sea interface, where the specialized fauna of Antarctica are concentrated and are most accessible for study. It is on a volcanic

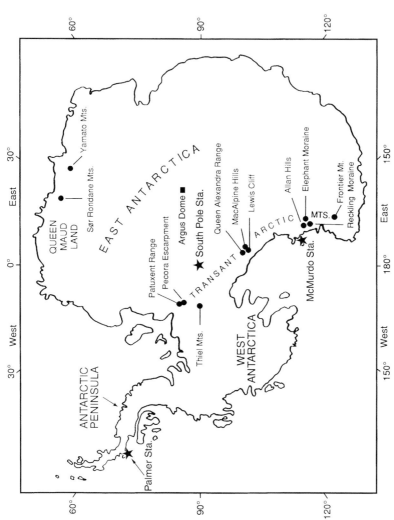

FIGURE 1.1 Map of Antarctica, showing the United States research stations (indicated by stars), the Transantarctic Mountains and some of the major meteorite concentration sites (indicated by filled circles).

island, with an active volcano whose lava lake and associated igneous rocks are of great interest to volcanologists. It is close to that part of the Transantarctic Mountain Range where the Dry Valleys are located. McMurdo has commodious laboratory facilities, extensive computer capability, and good communications with the outside world.

Many scientists operate out of McMurdo directly to nearby research locations. Cold adaptations and the metabolic effects of low temperatures on a wide variety of organisms, from penguins and seals, to fish, krill and bottom-dwellers that will not freeze when their body temperatures reach 0 °C, to the plankton on which they all depend, can be studied from nearby sea ice or with short trips along the shore. The Dry Valleys, only 60 km away, have a poorly understood microclimate that keeps them snow-free all year. The biology and geochemistry of meltwater lakes in the Dry Valleys also are not completely understood. Rocks are exposed in the Dry Valleys, and the geologist can study them there, unhindered by mantling ice. Fascinating adaptive responses to extreme conditions are displayed by endolithic organisms found along the very edges of the ice sheet. These are algae and fungi, living in symbiosis below the rock surfaces in a very special microenvironment beneath transparent minerals. They are able to spring to life almost instantaneously when the greenhouse effect of the overlying transparent mineral grains enhances the heat of a low, weak sun and when, simultaneously, liquid water becomes available. They can sink into dormancy just as rapidly when conditions change.

McMurdo-centered science also includes satellite-coupled meteorology, upper-atmosphere research and ozone-hole observations.

For those scientists needing access to more distant research sites, McMurdo has the capability to construct and support temporary remote stations for periods of years at sites where there is common interest from groups of scientists who wish to carry out a variety of research projects with greater than a one-year duration. For smaller operations, individual field parties can be put in to deep-field camps for periods of weeks, over a large part of the continent. McMurdo

Station is only three hours flying time away from the south pole, so it can support that very remote facility by air.

AMUNDSEN–SCOTT SOUTH POLE STATION

Research carried out at the geographic south pole, at an elevation of 3000 m in an exceptionally dry atmosphere includes meteorology, seismology as part of a global network of seismological stations, climatology with studies of snowmass accumulation trends and atmospheric trace constituents and aerosols, upper atmosphere studies, magnetosphere observations, cosmic-ray studies and, during the darkness of the austral winter, astronomical observations over significant fractions of the electromagnetic spectrum. A recent development is the use of the 3-km thick ice sheet as a highly transparent medium within which to observe, count and categorize the tiny flashes of Cherenkov radiation produced by neutrinos that have *passed through the earth.*

On a more mundane level, but more closely related to my interests, South Pole Station has become a collecting site for ancient cosmic dust particles.

PALMER STATION

Palmer Station is located on the Antarctic Peninsula, just north of the antarctic circle, and has subantarctic floral and faunal assemblages. Like McMurdo, it is located at the land–sea interface and much of the research carried out there involves marine ecosystems. This can be done in combination with research vessels, which find it much easier to visit Palmer than McMurdo. Tourist vessels also can reach Palmer, however, and there are ongoing programs to assess ecological damage due to tourism. Palmer has a seismic station as part of the global network, there is air sampling for trace gases and aerosols, and a range of upper atmosphere studies, including ozone-hole measurements linked to effects on marine microorganisms.

Palmer Station is run in complete separation from the much larger McMurdo–South Pole complex. Visitors to Antarctica via

McMurdo and South Pole Stations arrive mainly by air from New Zealand, while those arriving at Palmer have come by ship from South America.

LOGISTICS

In most NSF-supported, *non-polar* research, grants are made to the home institution of the grantee and, if logistics arrangements are necessary, funds are included in the grant for field vehicles and field equipment. For *non-polar* research, the grantee typically makes all his or her own arrangements. Because antarctic research can be quite dangerous, and because just getting to many remote sites is very expensive, the NSF's Office of Polar Programs is intimately involved in every aspect of the field work carried out on the continent, and runs an extensive operation that involves air transportation within Antarctica, the operation of research vessels around the coast, support of major research stations, supplying equipment, food, and clothing to scientific investigators and support of smaller-scale logistics needs such as snowmobiles and sledges for moving remote field camps. In 1976, which was the first year of my experience in Antarctica, US Navy personnel had a prominent role in the logistics operations, piloting and maintaining a fleet of six ski-equipped LC-130 Lockheed Hercules cargo planes and six Vietnam-era UH1N "Huey" helicopters. These helicopters ("helos") have an effective operating range of 185 km and the LC-130s can reach any point on the continent, if need be. Ski-equipped Twin Otters have been phased in in recent years to help bridge the gap between cargo-carrying capacities and ranges of the helos and LC-130s. Twin Otters are operated by Canadian bush pilots.

With shrinking military budgets, the Navy has been phasing out its participation in the antarctic enterprise. The 1995–96 austral summer saw the last of the Navy helos – they have been replaced by civilian contractor helos and pilots. Since the 1996–97 season, the Navy LC-130s and their pilots have been replaced by LC-130s and pilots of the New York State National Guard. At the same time, a civilian contractor ran many functions of the permanent US bases, and the Division of Polar Programs[1] oversaw the entire operation, with offices at McMurdo Station. So being funded for antarctic field work is not like any other geological field work, where you are on your own. In McMurdo, you become part of a rigid structure with a complicated hierarchy of procedures, requirements and rules. In this system you may not know immediately the best direction from which to approach a problem. It is always a great relief to escape this atmosphere to a deep-field camp on the ice plateau of East Antarctica, where survival may be more difficult but life is simpler.

A US-run deep-field camp in Antarctica is an interesting mixture of modern technology and a retreat to the past. We live in Scott tents, so named because they are designed after the tents that Robert Falcon Scott used in his polar expeditions early in the twentieth century. We do travel by snowmobile, which is more convenient than man-hauling sledges, as Scott did disastrously, or using dog teams as Amundsen did very successfully. We tow all our equipment, tents, food and fuel on Nansen sledges, designed in the nineteenth century after sledges used by eskimos. Modern touches, however, are appearing. A great convenience for mapping has been the introduction of Global Positioning System (GPS) instruments that can tell you where you are within a few meters on the earth's surface by triangulation, using signals generated by satellites. Lap-top computers are appearing in tents, and batteries are maintained in a fully charged condition by solar panels.

IF YOU WANT TO GO THERE

According to the Antarctic Treaty, the continent is reserved for scientific research, so to go there you should be a scientist or a science support person. Minor exceptions can be Members of Congress, selected

---

[1] The Division of Polar Programs, in 1994 became the administratively more important OPP, or Office of Polar Programs, in recognition of the increasing importance of Antarctica as a research site.

persons in the arts, selected newspersons, the occasional boy scout or girl scout and, on certain anniversary celebrations, OAEs (Old Antarctic Explorers). There are a lot of OAEs around. For the most part, they seem to cling to life with tenacity and zest.

Suppose you plan to go to Antarctica, as I did. At least six months before departure I had to have a complete medical and dental checkup. The NSF does not want personnel going to very remote sites, or even to McMurdo Station, with medical or dental problems that might require an emergency evacuation. The dental examination also has a grimmer aspect that is discussed among the grantees but very seldom by the grantor – dental records are a last resort for identification of very badly damaged corpses.

All personnel who expect to winter over also must have a psychiatric examination. This is supposed to screen out all those who are certifiably insane, but sometimes does not. There are those, of course, who claim that you have to be insane anyway to want to winter over. Now that I think of it, there are those who claim you have to be insane to want to go to Antarctica at all. Actually, the great majority of people can be divided into two groups of unequal size – a large group who would rather die than go to Antarctica, and a smaller cohort who would kill to get there.

In 1976, when I first started going to Antarctica, the results of the physical exams were forwarded to the ranking US Navy doctor who was to be in charge of the medical service at McMurdo Station. He had the right of final approval of your visit. When the Navy doctor found me to be fit, I knew I had an excellent chance of actually making the trip.

2   How the project began

ANTARCTICA AS A PLACE TO SEARCH FOR
METEORITES? YOU MUST BE KIDDING!
The concept followed no evolutionary path. It was suddenly there, as
bright as the comic-strip light bulb that signifies a new idea: *meteorites are concentrated on the ice in Antarctica!* The occasion was the
thirty-sixth annual meeting of the Meteoritical Society, which took
place during the last week of August 1973 in Davos, Switzerland. I was
listening to a paper by Makoto and Masako Shima, a Japanese husband
and wife team who are both chemists. He was describing their analyses of some stony meteorites. These specimens were interesting to
me because they had been recovered in Antarctica. The pre-meeting
abstract of the paper mentioned four meteorites that had been found
within a 5×10 km area, lying on the ice at the Yamato Mountains (see
Figure 1.1). I was quite aware of how rare meteorites really are, and
as far as I knew, when meteorites are found near each other, as these
had been, they are invariably fragments of a single fall. This was my
assumption in the present case, and I had attended this presentation
because of a long-standing, general interest in Antarctica, rather than
a specific interest in the meteorites to be described. Actually, the abstract made it clear that these specimens were of distinctly different
types, but I had been skimming and had not read that far. The key
word, so far as I was concerned, had been *Antarctica*.

It took some time to get used to Dr. Shima's accent, and it
was about halfway into the talk before I suddenly realized that he
was describing meteorites of four different types, and these could
not have come from anything but four separate falls. That is when
the light bulb went on over my head, and I thought, "Meteorites
are concentrated on the ice in Antarctica!" I suddenly started paying

very close attention. Repeating it over and over to myself, however, was not encouraging because I could think of no mechanism that would concentrate meteorites, much less one that would be unique to Antarctica. These insights would come only slowly, and much later. Clearly, also, I cannot claim to be the first to realize that meteorites must be concentrated somehow on the ice. Yoshida and colleagues in a 1971 paper that included as a coauthor Renji Naruse, the discoverer of the first Yamato Mountains meteorite, had already recognized the fact and they were wrestling with possible causes of the concentration process. Their general suggestion was that the meteorite concentration was related somehow to the movement and structure of the ice, and also that other such concentrations might be found. But I had not yet seen this paper.

After his talk at Davos, I spoke to Dr. Shima and he mentioned that the glaciologists had actually found *nine* meteorites in this small area, and superficial examination suggested that they were all different. So there *must* be some kind of concentration mechanism. Half an hour later, I started mentally writing a research proposal to visit Antarctica and search for concentrations of meteorites. The hypothesis was that the site where the Japanese scientists had found a concentration – the Yamato Mountains – could not be unique in a continent that occupied 9% of the total land surface of the earth. Others, however, were not so sure, and the eventual proposal, when submitted to the National Science Foundation's (NSF) Division of Polar Programs, was politely declined. It is easy to see why the reviewers were unimpressed, because aside from the apparently anomalous concentration of nine meteorites at the Yamato Mountains, only four other specimens had ever been found in all of Antarctica.

SOME HISTORY

The first antarctic meteorite ever found was a 1 kg L5 chondrite discovered during Douglas Mawson's Australian Antarctic Expedition in 1911–14. The distinction of finding this specimen belongs to an unnamed member of an exploration party led by Mr. F.H. Bickerton, whose mission was to explore and map westward from Mawson's base at Cape Dennison, in Commonwealth Bay on the Adelie Land coast. On their fourth day out, and only 43 km into their traverse, the three-man party found a meteorite, which they assumed was a fresh fall. The Adelie Land meteorite can be seen today at the South Australian Museum, in Adelaide.

Meteorites 2–4 were found subsequently at widely separated points in Antarctica: the second one, an iron meteorite, was found almost 50 years after Adelie Land in 1961 on a southern spur of the Humboldt Mountains by Russian geologists mapping near their base, Novolazarevskaya; Antarctica's third, a pallasitic stony iron in two pieces, was picked up in 1961 on ice in a moraine below Mt. Wrather in the Thiel Mountains by geologists of the United States (US) Geological Survey; and the fourth, an iron, was discovered in 1964 in the Neptune Mountains by geologists of the U.S. Geological Survey. At first glance, there was nothing to recommend the antarctic continent as a place where one could find great numbers of meteorites, since so few had been found.

With hindsight, of course, the following hints suggested the great potential of Antarctica as a meteorite recovery ground: (1) only a very small number of people had ever visited Antarctica, yet four of these had found meteorites; (2) only a very small total of the surface area of Antarctica had been examined on foot, yet this area contained four meteorites; and (3) one of the four meteorites, the pallasite, was of an extremely rare type, suggesting that many more, of the more common classes, should be recoverable. Singly, these hints are straws in the wind, but taken together they are somewhat suggestive. This perception is quite clear only now, but it received thunderous confirmation in events beginning in 1969 with the Japanese discovery of nine meteorites at the Yamato Mountains, and these events continue today. During only 20 years, the meteorites collected from perhaps 4000 km^2 of ice in Antarctica have doubled or tripled the number of individual specimens that had been accumulated in the world's museums over the preceding 200 years, collected from over 90% of the earth's land surface. Nonetheless, in 1974 we did not foresee what a magnificent place to collect meteorites the antarctic ice sheet really was, and the evolution of this idea and its implementation took a tortuous path.

THE JAPANESE CONNECTION

In 1973, Professor Takesi Nagata had been since 1961 a Visiting Professor at the University of Pittsburgh in our Department of Geology and Planetary Science. Typically, he would be present once or twice a year for periods of two to four weeks at a time, doing collaborative research with Mike Fuller and Vic Schmidt, who were professors in our department. Nagata was also the director of the Japanese National Institute for Polar Research. He always claimed that his time in Japan was completely occupied in administration, and the only times in which he could do any research were those short periods that he could spend with us, in the relaxed atmosphere of our department. In Japan, Dr. Nagata was always addressed in terms of the deepest respect, as Nagata-*san*. He encouraged us, however, to call him "Tak," more in line, I guess, with the American style.

During the fall of 1973, Tak came for a visit and I took advantage of the occasion to mention the remarkable meteorite concentration his people had found on the ice in Antarctica. Apparently nobody had told him about it. Here was a scientist who had been designated by the Emperor as a National Living Intellectual Treasure (this is no joke). I told him what I knew, and a little light immediately came on over his head, except that his was more appropriately a Japanese lantern (this is a joke). His thought was, "Meteorites must be concentrated on the ice in Antarctica!" My thought exactly, and arrived at with about the same speed – we had had a meeting of minds. We agreed on the fundamental importance of this concept, and I drew strength from the fact that one of us was a National Living Intellectual Treasure and also one of the few non-US members of the US National Academy of Sciences. He turned away to send a few quick telegrams back home asking for full details on the discovery and, after a few replies, sent a telegram to his field team in Antarctica, which was even then preparing for another visit to the Yamato Mountains, instructing them to search for meteorites. I returned to writing the unsuccessful proposal I mentioned above, to search for meteorites out of McMurdo Station, on the other side of the continent from Syowa Base.

We agreed to keep in touch on the matter, and so I learned eventually that his field team, pretty much in their spare time, had collected 12 more specimens during the December 1973 to January 1974 field season and, remarkably, these were all recovered at about the same place as the 1969 finds. Encouraged by this, I resubmitted my proposal with this new information. That season, during December 1974 and January 1975, the Japanese field team made an all-out effort and recovered a stunning 663 meteorite specimens at the Yamato Mountains!

At that time, Dr. Mort Turner was Program Manager for Geology at the former Division of Polar Programs (now known as Office of Polar Programs), and I had gotten to know him in the course of events involved in my unsuccessful research proposal. In an agony of frustration, I called him up and gave him the latest news. After only a moment, he said in a thoughtful tone of voice, "Well, the panel has just declined your proposal again, but they did not have this information. I urge you to resubmit it immediately, and I think it will be funded." And that is the way it turned out: we were funded for the 1976–77 summer field season, on the third try. The project would become known as the Antarctic Search for Meteorites, or ANSMET. The Japanese, meanwhile, recovered 307 more specimens during the 1975–76 austral summer.

When Tak next came to visit, I hastened to tell him that my proposal had finally been accepted, and that I expected to go to the ice in the 1976–77 season. He congratulated me, and then shocked me with, "Bill, I am planning to send a man to McMurdo that season also, to search for meteorites." I learned then that the Japanese Antarctic Research Expedition (JARE) apparently had had a cooperative arrangement with the US program for a number of years, and that Tak was sending his man as part of that agreement. I didn't know how to deal with this news, because my supposition had been that the Japanese would continue their very successful meteorite-collecting activities at Yamato Mountains, where they had searched only a small fraction of the icefields. Instead, they planned to suspend that operation for a while. Suddenly, I had a disturbing mental picture of field teams from two different countries competing for logistics support, competing to be the first to find meteorites in this place across the continent from the Japanese site, and competing to collect more meteorites than the other group. I waited to see if Tak would suggest some kind of arrangement to mitigate the seemingly destructive aspects of what he was planning, but he did not. I was too stunned to think creatively, so in deep confusion I let the moment pass. As succeeding months became busier, I was able partially to ignore this situation.

STARTING FROM SCRATCH

When I was planning our first field season, I had no clue about how the system operated, and I badly needed expert advice. I knew there was a dynamic group of polar scientists at The Ohio State University in Columbus, and I got the name of David Elliot, who was part of that group and at the time was also Chairman of the Department of Geological Sciences. I cold-called him, introduced myself, explained my situation, and he invited me over for a chat. David is a remarkably gracious person, and gave no hint of how I must have been disrupting his busy schedule. During the major part of one day, he explained exactly how the system works, what clothing and equipment I might wish to take along to supplement what would be issued, what materials I would not need to take, what the danger signs were in the field in a changeable weather situation, and something of the geology along the Transantarctic Mountains. This was my first introduction to anyone involved in polar research, and it was a very happy one. I later learned that these characteristics of friendliness and helpfulness are almost universal among principal investigators in this field; I want to call attention here to this noteworthy circumstance.

At about the time when I needed to decide on the makeup of a field party, I received a letter from Ed Olsen, Curator of Minerals at the Chicago Museum of Natural History. Here, suddenly, was someone who had been thinking about the Japanese antarctic meteorite discoveries and had decided on his own that the concentration at Yamato Mountains could not be unique. He had written to a colleague, Carleton Moore, in the Geology Department at Arizona State University, proposing the idea and suggesting they write a proposal to search for meteorites in Antarctica. Carleton apparently had recently reviewed my third proposal and had to tell Ed that he was too late. They are good friends; Ed finally got him to utter my name, and wrote, asking to be in the field party. I knew Carleton but not Ed, so I called Carleton and asked him how he thought Ed would be in an isolated field camp, under stressful weather conditions, when we might be tentbound for long periods. Carleton gave him his unqualified support, so I had a field partner.

The US Geological Survey has a tremendous library of aerial and satellite photos of Antarctica, and I became a frequent visitor to their archives, located in Reston, Virginia. The satellite photos were on filmstrips that could be projected for viewing. They were filed according to the geographic coordinates of the image, so when I wanted to search a certain part of the continent I would request whatever coverage they had within certain bounds of latitude and longitude. A technician would type those coordinates into a computer, which would relay them to a really big computer at the EROS Data Center in Sioux Falls, South Dakota, and this machine would then flash back a message telling the technician where to find the appropriate film spools on the shelves behind him. In this way, I was able to search large areas of the ice sheet by remote sensing. I looked not for meteorites, because they would be too small, but for large patches of ice because one wouldn't expect to find meteorites in deep snow. I learned that many very large exposures of ice exist in association with the Transantarctic Mountain Range, a magnificent chain of mountains that spans the continent. These mountains pass about 100 km west of McMurdo Station (see Figure 1.1).

When I had found some ice patches within helicopter range of McMurdo, I went to the US Geological Survey aerial photo library at Reston for a closer look at them. Presiding over this valuable scientific resource was an invaluable natural resource named Bob Allen, who had worked with every photo in the collection and seemingly remembered each one, in great detail.

In aerial photos one can often see details that indicate ice-flow directions. For example, lateral moraines are trains of rocks that have been scraped off the sides of mountains by a glacier and carried along until it joins another glacier, with its own lateral moraines. At the junction point their lateral moraines merge and are carried downstream after that as a medial moraine. By this pattern of moraines, one can learn the direction of ice movement. Tracing it backward, upglacier, one can determine the pattern of flow off the ice sheet as it sheds ice toward the edge of the continent. Ice acts as a very viscous hydraulic system, and when it encounters a barrier to flow, such as the Transantarctic Mountains, it can sometimes overwhelm it and flow over the impediment. In other places – the southern Victoria Land region opposite Ross Island is a good example – the mountains are too high and the ice is constrained to funnel through passes between the mountains. When this occurs the flow speeds up, as in any hydraulic system. Glaciers therefore tend to move more rapidly than the ice sheet that is supplying them. If ice accelerates it tends to form crevasses that can be seen clearly on aerial photos. Since acceleration also occurs when ice changes direction, crevasse patterns can tell much about ice flow at the edges of the ice sheet, where it is funneling into mountain passes. Specifically, one can see that there are some areas where a mountain barrier is long enough so that ice located near the middle of the barrier has very little chance to flow around it to get through the valleys on either side. These patches of ice appear to be stagnation points, where flow is not occurring and where any meteorites on the ice would not be swept down the valleys, and out to sea. I found a number of such areas within helicopter distance of McMurdo, and these seemed good places to begin our search.

GETTING THERE IS HALF THE FUN

First, of course, we had to get there. The NSF was using the Military Air Transport System (MATS) to ship its scientists and support personnel across the Pacific to Christchurch, New Zealand. I use the word "ship" because these C-141s were cargo planes, and we were treated pretty much as cargo. The difference was that we could load and unload ourselves, and wanted to, rather than be strapped to loading pallets and shoveled aboard by a forklift. The only other difference was that they had to supply toilets and box lunches for this particular cargo. Around the middle of November 1976, we marched on board and found chairs bolted to the floor in numberless ranks, where we sat, cheek by jowl, for a very long flight punctuated by refueling stops at Honolulu and Pago-Pago. The punctuations both occurred during nighttime hours, but it was difficult to get any sleep anyway, so they were welcome chances to walk around.

Cargo planes do not have much sound-deadening insulation. In a letter home, I described the flight as perhaps similar to flying sitting backward in an enlarged garbage can, with only a couple of portholes, while maniacs outside were beating the can with sticks. After about 22 hours we arrived in Christchurch during the early morning to find we had skipped a day, having crossed the International Dateline at 180° longitude.

In later years the NSF logistics people realized they could ship their people to New Zealand more cheaply on commercial airlines, availing themselves of government discounts. This offered a number of advantages, including the fact that grantees arrived with smiles on their faces.

New Zealand is a lovely country. Christchurch, a beautiful, smallish city on the South Island of New Zealand, seemed to be more British than Great Britain. We arrived there in late November, which is early summer in the southern hemisphere, and I was impressed by the gardens in front of private homes lining the road along which our taxi traveled. Even more pleasing was the presence of many old, stately copper beeches, one of my favorite trees. In later visits to Antarctica, I always looked forward to this first view of New Zealand.

On Oxford Terrace, in Christchurch, close by a small river, the Avon, stands a statue of Robert Falcon Scott in a heroic pose, dressed in canvas windbreakers, gazing intently and with determination toward (presumably) the South Pole. Scott occupies a mystical niche in the British psyche, but in no place more so than in Christchurch. It was from the nearby seaport at Lyttleton that Scott's last expedition set out, and Christchurch has always felt particularly close to his tragic enterprise. The entire story has been told many times, from many different viewpoints, both eulogizing and condemning him. An example of the eulogy is found in Apsley Cherry-Garrard's 1922 book *The Worst Journey in the World*, which is first and foremost a wonderful, powerful history. There are rumors, probably not true, that it was actually written by George Bernard Shaw, a next-door neighbor and good friend of Cherry-Garrard's. It is reissued periodically in paperback form, and anyone who appreciates excellent authorship and a gripping story should read this. Many years after Scott's death in Antarctica, excerpts from his recovered diary and the personal diaries of some of the surviving expedition members were made public. These made it seem that he had alienated many members of his expedition and had committed a number of mistakes in judgment whose cumulative effect brought death to him and his entire party of five. Thus it is possible that history had been blatantly rewritten in the interest of preserving Scott's memory as a hero. Interested readers can find these books also. Without offering any opinion on these extreme claims, I have to assume that Scott seriously overreached his abilities, and died as a result. Every time I visit Christchurch I spend a little time gazing at Scott's statue. Then I sit down somewhere and think about this, resolving not to do what Scott did.

The NSF maintains a warehouse at the airport with cold weather clothing. After a few days to rest up and explore, we gathered at this warehouse and were issued the clothing we would wear in Antarctica. These items are cleaned and repaired after every field season and reused by others the following year. When it came time to depart, we gathered again at the warehouse, removed our normal street clothes and put on our antarctic gear. Then we lined up at the scales to be weighed with our luggage, lined up to get on a bus that took us out to the plane, and lined up at the plane in the bright summer sunshine, sweating in our antarctic clothing, to receive box lunches and file on board. There were 50 or 60 of us geologists, geophysicists, biologists, meteorologists, meteoriticists, glaciologists, support persons and assorted other folks, all intent on making the most of our short summer season. All the scientists were hoping to make discoveries that would shake the scientific world. We were only 3000 km away from McMurdo, and on our way!

The C-141 Starlifter is a jet plane with four engines and, sitting on the tarmac, wings that appear to droop. It can make the trip to McMurdo in about 5:30 hours but can be used only in the early part of the summer season, when longer runways on the sea-ice are available for wheeled-aircraft landings. The other plane used for round trips between McMurdo and Christchurch is the LC-130, a somewhat smaller, propellor-driven cargo plane with four engines that requires eight to nine hours for the trip. Both have only a few windows; cargo doesn't have to see what is happening outside. On this first trip, we were lucky enough to be carried down on a C-141, so the trip was substantially shorter.

One hears a lot about the PSR (Point of Safe Return) on these trips. For each airplane this is a point beyond which it must continue on to McMurdo because it no longer has enough fuel to return. Because the weather at McMurdo can turn nasty very, very rapidly, and without much warning, pilots are always in touch by radio as they approach their PSR, preparing themselves for the commitment to continue on, or the decision to turn back.

Often, flights will turn back because of sudden weather changes at McMurdo. Bob Fudali, who has been a team member on a couple of our expeditions, recalls such a flight that also carried a group of New Zealand scientists, popularly known as "Kiwis." Apparently this group had been partying to some extent (possibly to a great extent) the night before. They boarded the plane in what could be described as somewhat weakened spirits and promptly went to sleep. Three hours into the flight, we learned that we were turning around and heading back to New Zealand. The Kiwis missed this announcement, which had come over the loudspeaker in a pretty garbled form. During our approach to Christchurch airport they woke up, just in time to put on their cold weather clothing and emerge onto the tarmac in full windpants and parkas in the middle of summer.

On my first flight down in 1976, shortly after we had passed our PSR, we began to see Antarctica. We flew parallel to, and east of, the northern end of the Transantarctic Mountains. Of course, since this is a transcontinental range, it has *two* northern ends – it could only happen in Antarctica! Taking turns at the windows, we saw peaks blanketed by snow and valleys filled with it, so that all the sharp edges had been rounded. The scene looked more lunar than terrestrial. I remember my illogical reaction at finally seeing Antarctica for the first time: "By God, it actually *is there*!" In another hour, or so, we landed on the ice of the Ross Sea, between McMurdo Station on Ross Island and the Transantarctic Mountains on the antarctic mainland.

THE BIGGEST LITTLE TOWN IN ANTARCTICA

McMurdo Station is a collection of buildings reflecting a number of different eras of construction and aspects of age, planted on the edge of Ross Island. The island itself is volcanic, with two major volcanoes, one dormant and one active. These are Mts. Terror and Erebus, respectively. Mt. Erebus, named after the head hellhound himself, usually has a plume of smoke or steam issuing from the crater at its 3795 m summit, and in the crater is a lava lake that attracts volcanologists like moths to a flame. So far, however, none have been burned. The island is covered deeply in black volcanic ash from prehistoric eruptions. This gives McMurdo a pretty dismal appearance, which is heightened during the summer months as much of the snow melts and more of the land surface is exposed. This US base is located at 77.5° south latitude, still 12.5° from the geographic south pole. LC-130 flights from McMurdo to the south pole take about three hours.

In the summer season, McMurdo hosts about 1200 people. Very few individuals are seen on the streets, except during meal times, when everyone goes to and returns from the mess hall. Aside from the occasional need to walk between buildings, there seems to be nothing in the uninspiring surroundings to draw one outside. The scientists usually spend as little time as possible in McMurdo, preferring the stimulating environments of the areas where their research is located and where one can savor the true character of this continent. The support people, on the other hand, are generally stuck in McMurdo as mechanics, medical corpsmen, dentists and doctors, carpenters, electricians, plumbers, mess hall cooks and servers, secretaries, equipment specialists and even janitors. Many of these people signed-on with the idea of finding challenge and adventure on the "White Continent," but find themselves instead carrying out routine tasks on a black island. A specially favored class among the support personnel are the helicopter and LC-130 pilots and crew, who get to carry the scientists around a significant part of the continent to their research areas. Like the others, however, they pretty much stay inside when they are in McMurdo.

THE JAPANESE RECONNECTION

When Ed Olsen and I arrived at McMurdo in November of 1976, we learned that Takesi Nagata and two geologists from the Japan National Institute of Polar Research, Keizo Yanai and Katsutada (Katsu) Kaminuma, were already there and had made a helicopter flight up the Taylor Glacier, searching for meteorites. They had not found any, but I was once again upset by the uncompromisingly competitive nature of what Tak proposed to do. Apparently there was now also some concern about this at the Chalet (the A-frame containing the administrative nerve center at McMurdo), and as a result the Chief Scientist at McMurdo, Duwayne Anderson, proposed that our field parties be combined. Tak said he would agree to such an arrangement, and my anxiety began to subside. Duwayne asked me to draw up an agreement that was acceptable to both Tak and me, for us both to sign. I got Tak's thoughts on the matter and discussed mine with him, then wrote a simple statement saying that we would cooperate in meteorite searches and divide equally any meteorites we might find. We both signed the statement and gave it to Duwayne. This agreement solved a lot of potential problems and heralded many future successes. The text of the agreement is included as Appendix A.

As meteoriticists, we owe much to the Japanese. We are in debt to Renji Naruse, specifically, for recognizing the first Yamato meteorite. We are in debt to the Japanese meteorite collection program for adding thousands of new meteorite specimens to the world's collections. We are in debt to our Japanese field collaborators over the years: Keizo Yanai, Kazuyuki Shiraishi, Fumihiko Nishio and Minoru Funaki for their friendship and for teaching us how to survive on the ice plateau of East Antarctica. I am personally in debt to the Japanese program because, without their results, my modest proposal never would have gotten past the reviewers and never would have grown into the US-sponsored Antarctic Search for Meteorites (ANSMET) program that continues today, and that to date has contributed so many new meteorite specimens to the world's collections; all available for scientific research.

FIELD PREPARATIONS

Because Katsu had other research activities at McMurdo and because Tak had a heart condition that would not allow him to spend any time up on the ice plateau, where we would camp at an altitude of 2000 m, we became a three-person team.

We now had to get to know Keizo (pronounced *kayzo*) Yanai. We discovered that he had extensive antarctic experience and had been part of the team that collected meteorites at the Yamato Mountains, so he had "caught meteorites in the wild," so to speak. Language was a bit of a problem. In common with almost all other Americans, we spoke no Japanese. We had much difficulty communicating with him in English, even though he tried very hard, with frequent use of a dictionary. Keizo drew up a suggested grocery list for our field provisions. In checking through it, Ed puzzled and scratched his head for quite a while, finally asking me, "What is frozen brocori?" We realized that Keizo wrote phonetically, according to the way he would pronounce a word. We gradually learned the rules for conversing with Keizo. We were careful to pronounce words clearly. He improved a lot during the field season, but he also had an effect on us. Early in the season we were to work next to a small peak called Mt. Fleming. I still prefer to call it *Mt. Freming*. It just sounds right, somehow.

When we set up a camp, it had to be transportable by helicopter. In the early years of the ANSMET project, we were limited in the sites we could explore by the 185 km range of the helicopters. This limit is expanded slightly toward the northwest due to the presence of a refueling station on the mainland at Marble Point, 90 km distant from McMurdo Station.

I had made some preliminary plans for the field work, and Keizo was content to go along with them. In the Transantarctics, westward across the sound from McMurdo, was a region called the Dry Valleys (see Frontispiece). This is a group of glacially carved valleys that no longer have glaciers in them, and freshly fallen snow in these valleys sublimes away before it can accumulate. Apparently there is a microclimate here that keeps the valleys always snow-free, thus the term "dry." I had noticed a tongue of ice that descended from the ice plateau into one of these valleys – Upper Wright Valley. The ice came down off the plateau and traveled perhaps 2 km before sublimation and outright melting terminated it. Surely, here was a conveyor belt that had been carrying meteorites off the ice sheet for millions of years and depositing them in the moraine at its end! We would camp on the bare rock below the terminal moraine and search the moraine for meteorites. For a second camp we could spend a couple of days on a small ice patch 1000 m higher, between Mts. Fleming and Baldr.

SUCCESS! SUCCESS!

Ed, Keizo and I were ferried up to our first campsite in a helicopter piloted by Lt. Sam Feola. Sam is a Vietnam War veteran who combined great flying skill with an inquiring mind and tremendous enthusiasm for the experience of being in Antarctica and helping in scientific research. His co-pilot and crewman on this trip apparently also shared his enthusiasm. When we had unloaded our gear and supplies below the end of the Upper Wright Glacier, Sam calculated that he had some extra time that we could use for reconnaissance, if we wanted to. I asked him to fly us up to the planned site of our second camp, to see if it was a suitable spot. To get there, we flew 10 km while the surface below rose vertically through 1 km. Most of the vertical ascent was over a spectacular icefall cascading down through a feature called Vortex Col. At the top was a small patch of exposed ice about 3 × 3 km in area, partially covered in a very patchy way by snow. It didn't look very promising, but we got out and Keizo almost immediately spotted a meteorite near the helo. We were ecstatic, but while we were photographing it and admiring it, Keizo was scanning the ice with his binoculars. Presently he started running and we, of course, followed. Sam ran back to the helicopter and took off, following us at an elevation of about 2 m, and 20 m behind. In silhouette, it would have been a memorable, if puzzling, tableau: a figure running at top speed over the ice, two more people chasing him and a helicopter skimming along behind. I remember looking back and thinking that this is what an insect must feel like, being stalked by a praying mantis. Meanwhile, Keizo had spotted another meteorite.

So Keizo Yanai had found our first two meteorites during our first 20 minutes in the field, and that is how it all began. We didn't find another for the next six weeks!

3 The first three years

We had lost a lot of time getting started. I was already 49 years old before I had ever been to Antarctica. Two years that I could ill afford had been wasted in fruitless attempts to convince skeptical reviewers that meteorites occur in concentrations on antarctic ice. This is not a condemnation of the system: I can always be convicted of writing unconvincing proposals.

The system we have in the United States (US) for deciding whether or not to grant funds for new research is probably the best that can be designed. Major granting agencies like the National Institutes of Health (NIH), the National Aeronautics and Space Administration (NASA) and, specific to this case, the National Science Foundation (NSF) invite research ideas from the general community of scientists. These ideas are written in the form of research proposals, which are then sent to three or more reviewers in the same field as that of the proposer. In a majority of cases, the reviewers actually know the proposer and, in order to preserve friendships if possible, they are protected by a cloak of anonymity. This is a wise provision, but very frustrating to the proposer who receives bad reviews and wants to seek out these less insightful colleagues and give them a good shake!

In the case of my proposal to search for meteorites in Antarctica, the reviewers apparently could not accept as significant the early evidence of the Japanese experience that to me seemed so clear. Finally, of course, they became believers when faced with the fact of 663 specimens found by the Japanese at the Yamato Mountains during one field season. Before this, apparently, the possibility was dismissed without much thought. Years later, in reminiscing, Mort Turner told me that the general consensus among the early reviewers had been, "If the

NSF wanted to send someone to Antarctica to look for meteorites, Cassidy would be a good choice. But he would not find any."

Proposal deadlines for work in Antarctica are due every year by June 1. This allows about six months for the review process and then about one year to prepare for the austral summer season, the best months of which are December and January. The proposal that was eventually successful, although it was declined initially, had been submitted on June 1, 1975 and was for work in Antarctica during December, 1976 and January, 1977. This grant had been for one year. More in hope than through foresight, I had submitted a renewal proposal before the June 1, 1976 deadline, anticipating success and allowing for the 18-month delay before the antarctic summer season for which funding was requested. In this proposal I believe I mentioned the speculation that ice flowing from the center of Antarctica had ample time in which to collect falling meteorites and deliver them to the edges of the continent, where they were conveniently revealed in places where the katabatic winds had stripped off the snow cover. Later, after talking to a few glaciologists I recognized this hypothesis as rather simplistic, but it was the basic assumption under which we operated during our first field season.

After our climactic first 20 minutes in the field, during which we found our first two meteorites on an exposed patch of ice near the edge of the continent, the hypothesis seemed to have been vindicated. The next six weeks, however, were not simply anti-climactic, they were seriously discouraging, and I began to worry about the fate of my second proposal. We visited many ice patches along the plateau side of the Transantarctic Mts., being lifted in by helicopter and camping for several days to a week at each site. Everywhere, it was the same. The two meteorites we had already found at our second campsite turned out to be the only ones in this very large area. I didn't understand the reason for this until 1980, when David Drewry, of the Scott Polar Institute, discovered a feature that came to be known as the Taylor Ice Dome. An ice dome is well described by its name: it is essentially a hill of ice. One doesn't necessarily know what causes ice domes, but it is a fact that they exist, and the Taylor Dome exists only a few tens of km inland from the ice patches we had been visiting. It measures about 30 km wide by 80 km long, and it diverts the ice trying to flow from the continental interior toward the Dry Valleys.

If you could draw elevation contours on the ice surface, you would find that ice flows down-contour. This means that the general direction of movement for ice is along the long, gradual slope from the high elevation at the center of Antarctica toward sea level. If this ice encounters an ice dome, it flows around it, rather than trying to flow uphill to get over it. Because of the Taylor Dome, ice coming off the ice plateau of East Antarctica divides to join major drainage patterns located to the north and south of McMurdo. So in working close to McMurdo we had been prospecting in a barren area, because ice from the tremendous meteorite-collecting areas in the interior was being diverted away, and the only meteorites to be found where we had been looking either had fallen directly onto those spots or had been carried to them from the very small collecting area between the Taylor Dome and the Transantarctic Mountains.

WE GO TO METEORITE HEAVEN

Based on the two specimens we had found, I had been prepared to proclaim the trip a success, but the key to real success turned out to be exploring farther north. We had been working our way slowly toward the north, exploring a series of ice patches without result, and finally spent a week near Mt. DeWitt, on the south side of a chaos of fast-moving ice called the Mackay Glacier. From our camp at Mt. DeWitt we had often gazed at a flat-topped mountain to the north that almost always had a pancake-shaped cloud lying on it. The map told us this was Mt. Brooke. I thought the cloud also was a permanent enough feature to be named. There used to be a character traveling through the "L'il Abner" comic strip with his own personal rain cloud over his head. His name was Joe Btsplk. As mountains go, Mt. Brooke was a Joe Btsplk analog.

We looked often to the north, attracted by the Mt. Brooke cloud, and could not fail to notice also a very large patch of ice a little to the west of Mt. Brooke. Between that ice patch and Mt. DeWitt was the Mackay Glacier, carrying ice down to the Ross Sea. This glacier is the outlet for polar plateau ice that had been diverted northward around the Taylor Dome. We didn't know it at the time, but every site we had visited until then was included in the "flow shadow" of Taylor Dome, and the potential for collecting meteorites at all these sites was very limited. Looking across the Mackay Glacier at the great sky-blue patches of ice beyond Mt. Brooke, we were looking for the first time at ice that had a tremendous upstream collecting area. We were looking at Meteorite Heaven, but we didn't know it.

The closest mapped feature to the ice patch north of the Mackay Glacier was a low-lying, roughly Y-shaped ridge of rock called Allan Hills. Back at McMurdo, when I requested a reconnaissance flight to Allan Hills I was told that it was not possible because it was outside the permitted range for helicopter operations. Chatting with some of the helicopter pilots later that day, however, I found one who had actually put in a field party at Allan Hills during the 1975–76 season. This information carried weight, so we received permission to carry out the recon flight and were able to arrange one for January 18, 1977. The helicopter had its usual crew complement: a pilot, copilot and crewman. The season was drawing to a close and we looked on this as our last opportunity.

People with good eyesight are described as having 20/20 vision. When I enlisted in the US Navy in 1945, at the age of seventeen, I had eyesight measured as 20/10. This meant that I could see at 6 m the same degree of detail that a person with average sight could see only at 3 m. I could read an automobile license plate a block away. Of course, this had changed a lot from the time I was seventeen. Our pilot for the recon flight to Allan Hills was LCdr. Mike Brinck, whose vision must have been 20/5! We flew out to the Allan Hills ice patch and, after a few minutes of cruising above the ice surface in a near hover, Mike set down next to a small rock. It was a meteorite! We found three more specimens during the next hour and returned to McMurdo triumphant. In a few days we were allowed another flight, this time with Lt. Mick Brown at the controls. Mick was the most overtly competitive helicopter pilot I have met during 15 years in the antarctic research program; he knew we had found four meteorites with Mike Brinck and he was going to beat that number if he had to run out of fuel doing it.

We quickly found two specimens. Hopes rose for the rest of the flight, but we saw nothing. Finally, Mick announced that we had to return, "and anyway, there was nothing but a scattered moraine up ahead." It seemed to me to be a strange place for a moraine, and I asked Mick to land so we could check it out. We all went in different directions, and soon we were all standing by a different rock, jumping up and down to signal the others to leave their rocks and come to see the meteorite we had found. It turned out that the "moraine" had no terrestrial rocks at all – only meteorites. These undistinguished-looking, yellowish-brown lumps with yellowish fracture surfaces were all pieces of a single large stone meteorite, the sum of whose components weighed 407 kg. To load the largest of the 34 fragments, Mick flew the helo over and landed right next to it. Three of us were then barely able to lift the specimen and deposit it on the cabin floor. Then we went home to McMurdo. I shall never forget that day.

HUMAN NATURE BEING WHAT IT IS ...

With the resounding success of our first field season, it became easy to justify further work. But success can breed problems, and we had our share. Word had spread through the small community at McMurdo, and when we had packed our specimen boxes, nailed them shut, addressed them and strapped them for shipment, they were easy to identify as ours. And anyone could know they contained meteorites. Who would not want this kind of souvenir from Antarctica? So it was that after we had left for home someone broke into one of the Japanese boxes and removed a number of pieces of the 407 kg meteorite. These apparently were distributed into several willing hands and the violated sample box was hidden in a dark corner of the Berg Field Center. When the collection arrived in Japan the box was missed, and Keizo immediately telegraphed a query as to its whereabouts. I got in touch with the DPP and they made a search at McMurdo, finding the box with some samples remaining in it. They had no hard evidence that anything had been taken, although the inference was clear. They closed the box and shipped it to Keizo, who then realized that fragments had been stolen. To add insult to injury, however, the remaining samples had not been repacked properly and had suffered mutual abrasion in transit, producing a lot of crumbs and arriving in a substantially degraded condition.

Months later, a principal investigator saw a meteorite fragment on his assistant's shelf. The assistant readily admitted it was an antarctic meteorite and claimed to have walked in on the looting of the box by people he did not know. This is the only one of the missing fragments we have ever recovered. The only lucky aspect of this unfortunate incident is that the missing fragments were only a small part of a much larger specimen. Other boxes that could have been opened instead of this one contained the complete Japanese portions of the other meteorites.

After our second field season we experienced a new problem. This one grew in the fertile ground of our success – other people wanted in. We had recovered a carbonaceous chondrite during the 1977 field season. This excited a lot of interest because carbonaceous chondrites typically contain complex organic compounds, some of which are found in living tissue as the building blocks of protein. This fact could have profound implications for the origin of life in the solar system, or even in the universe. Either that, or it could be completely trivial. At that time, we didn't know, and we still do not know.

Well, we had found a carbonaceous chondrite. This was the first of many, but we didn't know that then. NASA made a press release and science reporters wrote it up with gusto. The discovery caught the eye of Cyril Ponnamperuma, Director of the Institute for the Study of the Origin of Life, at the University of Maryland, and he submitted a proposal to search for meteorites in Antarctica. His proposal would be in direct competition with my proposal for continuation of the ANSMET project.

This was a stressful situation because I saw myself in competition with a scientist who was running an entire institute on multimillion dollar research grants. I was afraid that his prestigious scientific presence and proven grant-getting ability would count heavily in his favor with reviewers. At my request, we met for lunch in a small restaurant in Washington, DC. He turned out to be a very nice person who was so excited over the possibility that Antarctica would turn out to be a storehouse of carbonaceous chondrites that were virtually uncontaminated by the terrestrial environment that he would be willing to go there, at great personal inconvenience, to search for them. I liked him immediately, and suggested that instead of competing, he might wish to be co-principal investigator on my proposal. He would not consider this compromise, however, preferring to let his own proposal stand.

As it turned out, Polar Programs made a policy decision to fund only one meteorite-collecting project, and to continue with the one that had so far been successful.

PLANNING FOR THE FUTURE

After the first field season I was beginning to think about the wide community of interest represented by that group of scientists around the world who were actively engaged in meteorite research, and I was certain there would be great demand for antarctic meteorites for many kinds of research. With the interests of my university department in mind, I decided to propose a curation center for antarctic meteorites at the University of Pittsburgh. The idea was to distribute portions of recovered antarctic meteorites to researchers all over the world. As it turned out, however, larger moves were afoot and there were indistinct, but ominous, sounds as of leviathans lumbering about, only half-seen in the flickering shadows at the edge of the clearing.

What to do about the antarctic meteorites? This question became a preoccupation for personnel at NSF, who knew by now that it would become an important issue, and it became a rosy possibility for two other large organizations: The Smithsonian Institution and NASA, both of whom apparently had awakened simultaneously. They had been sleeping in the same bed, so close together that four fists were now rubbing the sleepy-bugs out of four eyes indiscriminately. A ponderous mating dance now began and prudence dictated that minor actors such as I stand aside, for

> "When elephants fight, the grass is trampled; but when elephants make love the grass also suffers."[1]

The eventual arrangements were as follows. NSF stipulated its ownership of all meteorites to be collected in Antarctica with NSF funding. NASA agreed to be curator for all such meteorites, and the US National Museum (Smithsonian) would be the final archival site for antarctic meteorites collected in the US program. Almost as an afterthought, the collecting program would stay, at least for the present, at the University of Pittsburgh. I drew a little diagram and called it The Tetrahedral Agreement (Figure 3.1). I used the diagram in talks at NSF and NASA, but no one appreciated the irony in the relative sizes of the depicted partners. Leviathans do not seem to laugh a lot.

Actually, the results of this interagency agreement have been demonstrably good. NASA uses exacting laboratory sampling techniques and distribution procedures patterned after their practice with lunar samples. A thin section from each meteorite is forwarded to the US National Museum for initial description by petrologists there. For years, Brian Mason, Curator of Minerals at the Smithsonian, did microscopic examinations and wrote thousands of initial petrographic descriptions of stony meteorites we had collected. These descriptions are mailed out on a regular basis to a worldwide clientele of scientists,

---

[1] Adapted from a traditional Masai proverb.

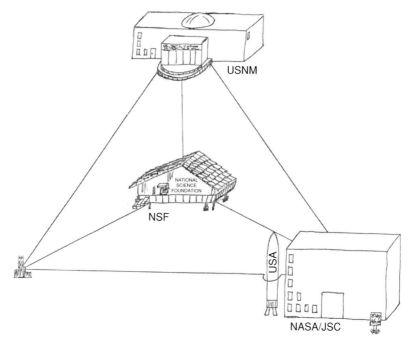

FIGURE 3.1 The "Tetrahedral Agreement" between The National
Science Foundation (NSF), symbolized as the headquarters chalet at
McMurdo, The National Aeronautics and Space Administration (NASA)
and the United States National Museum of the Smithsonian Institution
(USNM), with the University of Pittsburgh tacked on as a fourth partner.

who can then decide which specimens to request for their research.
When demand for a particular meteorite tapers off, it is transferred
to, and becomes the property of, The Smithsonian Institution. It can
then be distributed from there for further research if the occasional
request comes in.

NASA also supplies collecting materials to the field team, so the
specimen can be collected and transported in a way that avoids any
artificial contamination. Because there is often ice or snow adhering
to the meteorites, they are shipped back frozen to Houston, where
they are thawed out and dried rapidly in a stream of dry nitrogen.
This avoids leaching effects from liquid water during transport that
otherwise would compromise the specimens. NASA also supplies

special burglar-resistant shipping containers for transporting the samples from McMurdo.

There were major advantages that accrued to both the US National Museum and to NASA under this agreement. For NASA, missions to the moon had ended. Curation of lunar samples had been a scene of frenetic activity but was now settling down to a relatively constant, more routine pace. Their lab technicians were highly trained, highly skilled, intelligent workers who needed a periodic challenge. Antarctic meteorites supplied that challenge because a new collection would be arriving every year. For the US National Museum, the prospect was one of eventual great augmentation of their meteorite collections as the specimens were transferred from the curatorial stocks at NASA. There was even an advantage to me – I had some big new friends. Everything seemed to be going along pretty well. Faced with smothering love on the part of NASA and the US National Museum, I had given up the idea of proposing a curatorial center for the University of Pittsburgh and had admitted that they really could do the job much better. And so it was decreed.

The only thing we forgot to do was to inform Takesi Nagata about these arrangements. Our agreement had been for only one field season, but it now developed that Tak planned to keep sending participants for another two years. We decided to accommodate him in this, and it never occurred to me that shipping an entire specimen to Houston, frozen, for subdivision there would be anything but a desirable procedure in Tak's eyes. As it turned out, however, when Keizo returned after the second season without the Japanese half of the collection, he suffered enormous criticism for allowing this to happen. It took a face-to-face conference between Tak and the Director of the Division of Polar Programs, Ed Todd, to obtain Tak's reluctant agreement to this procedure. Even then, we had to promise not to circulate any preliminary descriptions before the Japanese had received their half of the collection. They also wished to be present when the collection was opened, and we solved this by inviting Keizo to Houston to be part of the curation process. We had firm friendships with the field people that Tak sent, but after the theft of some of the Japanese collection the year before, and our perceived expropriation of the 1977–78 collection, a lot of trust-building activity would be necessary now if we were to impress Tak with our continuing good intentions. There was a lot of debate over whether or not we wanted to exert all that effort.

OUR FIELD PROCEDURES EVOLVE

With the arrival of our second field season (1977–78), we had already a suspected high concentration of meteorites at Allan Hills and our mode of operation changed. In the previous season we had visited Allan Hills twice; each time for only a few hours. This time we went in with the intention to camp for some time at the site, collect as many meteorites as we could, and carry out reconnaissance flights only toward the end of the season to visit other nearby blue ice patches. Division of Polar Programs personnel had gotten past their previous year's nervousness about the distance to Allan Hills and were quite willing to allow us to fly in by helo. The pilots, of course, had always been eager.

Also, from now onward we would carry out field curation of collected meteorites, touching them only with tongs that had been specially cleaned at the NASA/Johnson Space Center. The specimens were stored in specially cleaned Teflon bags and sealed with very expensive tape. This last item was the result of an extensive search for tape that would retain its adhesive properties in extremely low temperatures; many were found that would not, and only one was found that would. Before collecting, each meteorite was to be described visually and photographed with a centimeter scale for size. Sometimes we had to examine a stone very closely to determine its meteoritic nature. We all suffer from drippy noses in cold weather, and we had to be particularly careful during a close look to avoid contamination of the meteorite from that source.

During this period we established our original discovery site at Allan Hills as a major concentration of meteorites, and reconnaissance visits to other ice patches in the general area also yielded specimens. We began calling the ice patch adjacent to Allan Hills the Allan Hills Main Icefield, to differentiate it from three others that we called the Allan Hills Near Western, Allan Hills Middle Western and Allan Hills Far Western Icefields. By analogy with our developing understanding of the field situation at the Allan Hills Main Icefield, we felt that if we could find two or three meteorites during a brief visit to a new site, then there must be many more that we had not found. All of these ice fields met that criterion.

Also during this period, a field party under Philip Kyle, who at that time had a postgraduate appointment in the Institute of Polar Studies at Ohio State, crossed an ice patch to the north of Allan Hills near Reckling Peak. During their traverse, they picked up five meteorites, which they presented to us after the field season. We hadn't yet visited the site ourselves, but the finds made by Phil's party seemed sufficient justification for including the Reckling Peak Icefield in our list of bankable assets.

One of our early challenges was in maintaining communications with McMurdo. All field parties are expected to check in by radio once a day, and for this we were given a military field radio, which ran on batteries. The batteries quickly lost their charge in the cold, and while we could hear McMurdo, they could not hear us. We learned to make marginal contacts for a while by heating the batteries ahead of time over the stove. Over the years, this situation improved markedly. We learned that nearby field parties could sometimes hear our transmissions and relay our messages. We also found that at times we could reach distant stations like South Pole because the signal bounced off the ionosphere, while the signal to a nearby station had to travel line-of-sight. Today, we use solar cells to maintain the charge in the batteries, and also take along a couple of spare radios, in case internal failures develop.

At the height of our communications problems during the second year, we had not successfully made contact for several days. There was some concern at McMurdo about this, so they assigned one of the LC-130 missions that was leaving that day to make a slight detour, to overfly our camp, and try to get us on the radio. This four-engine plane came in while we were still asleep and buzzed us at about 10 m above our tents. When it is that close an LC-130 makes a noise so shattering you can feel it; I woke up cringing in my sleeping bag, with nowhere to hide. When I had recovered my wits I stepped outside and talked with the pilot on a small hand-held radio that we had, and explained the problem with the batteries. He relayed the message, so we had a good check-in that day. Then the rascal buzzed us again and went away. I didn't see him leave because I was lying face-down in the snow.

FIELD CONDITIONS

After the first field season, Ed Olsen's boss at the Chicago Museum of Natural History saw no future for Ed in Antarctica. This is understandable, if regrettable, because Ed would have been placed in the position of collecting meteorites for the competition (i.e., the Smithsonian), so I had to find another field partner. During our second season we were a four-man team. My field partners were our Japanese contingent, Keizo Yanai and Minoru Funaki, and a long-time colleague, Billy Glass, Professor of Geology at the University of Delaware. We had taken a large army-surplus hexagonal tent to the field with us. In addition, we had a two-man Scott tent from McMurdo and a two-man Scott tent supplied by the Japanese program. We planned to use the Scott tents for sleeping and the big tent for storing supplies and cooking. Scott tents let in a lot of light, but the army tent had opaque walls, so it was very dark inside. Being hexagonal, too, the square ground canvas we brought did not fit exactly, so there were areas of bare snow inside the tent on two sides. Yanai and Funaki took it on themselves to prepare our first dinner in the field, and they disposed of some no-longer-wanted hot water simply by pouring it into the snow at the edge of the canvas floor. Suddenly a beam of lovely blue light dimly illuminated the interior of the tent! The effect was startling in its unexpectedness. The light was streaming upward from the melt-hole in the snow. Looking down the hole, we realized that the snow layer at this point was only about 20 cm deep over ice that was at least many meters thick. The hot water had melted the snow and exposed the ice surface. The sun was shining outside, 24 hours per day, as it does all summer in Antarctica, and sunlight was penetrating the ice. The ice is clear enough that visible light, which is composed of all the colors of the rainbow, has a long path length before being reflected back toward the surface. When it emerges it is predominantly blue because the ice has preferentially scattered and absorbed the other colors. It requires a long path length in clear ice for this effect to become noticeable, and antarctic ice tends to be very clear. We spent some time melting snow, heating the water and making blue floodlights along the edges of the tent, until it looked inside kind of like a nightclub in the middle of Antarctica.

One of our campsites during the second field season was located near Carapace Nunatak, which had large blue-ice areas nearby. Carapace Nunatak is a rock outcrop consisting of lava flows that had been extruded into a fresh water lake. The lake water had chilled the lava quickly, so that it was formed into frozen lobes called pillows, about 1 m in diameter. These pillows had cooled from the outside inward, and all residual liquids remaining, after most of the crystallization had taken place, had been concentrated near the centers of the pillows. The last stage of crystallization in these small cavities characteristically had produced lovely quartz crystals lining the insides of the cavities. These had weathered out of the pillows as masses called geodes, and fragments of them were to be found along two linear moraines that formed ridges of broken rock carried downstream by the ice from Carapace Nunatak. It was a strong temptation to spend odd moments searching for geodes along the moraine ridges.

Billy Glass, Keizo, Funaki and I were doing just that. Keizo and Funaki had moved up close to the nunatak, and Bill and I were prospecting the moraine ridges. The ridges were raised about 2 m above a snow street that separated them. The wind was blowing a near gale crosswise to the moraines, and our camp had been set up about 30 m beyond the end of one of the moraines. In the early years I always had some concern that our tents would blow away. We had progressed perhaps 300 m along one of the ridges when I looked back in the direction of our camp and was shocked to see the Japanese Scott tent with its taut canvas walls ballooned out in the wind. The shocking thing about this was that the nearby, larger hexagonal tent was not there, and the Scott tent from McMurdo was also missing! I called Bill and we descended into the snow street where fast walking was easier and hastened back in the direction of the camp. About 50 m further on I looked up and the Japanese tent had disappeared also! I was thinking then that we would have to salvage whatever we could find and might have to dig a snow cave in which to survive until the helicopters came looking for us. After another 200 m, however, we saw the peaks of all three tents sitting right where they should have been, in a hollow below the end of the moraine. Then I realized that we should not have been able to see *any* of the tents from my vantage point on the moraine, and I must have seen the one Scott tent in a mirage.

An unexpected phenomenon in Antarctica is the mirage. We would often see mountain peaks ahead of us, rising above the horizon long before we should have been able to see them. A mirage results when the atmosphere near the surface is less dense than an overlying layer of air. The wind rushing over the moraine ridges might have been less dense than the air above it, because it was constrained by the topography to travel faster. I was looking along the ridge when I saw one of the tents in a mirage. In retrospect, the tent also looked much closer than it should have, which is also consistent with the effects of a mirage.

The real devil in antarctic research is the wind. The following is an extract from *The Home of the Blizzard*, by Douglas Mawson[2]:

> Shod with good spikes, in a steady wind, one had only to push
> hard to keep a sure footing. It would not be true to say "to keep

[2] Mawson, D. (1915) *The Home of the Blizzard*, vol. 1, p. 116 ff. London: William Heinemann.

erect," for equilibrium was maintained by leaning against the wind. In course of time, those whose duties habitually took them out of doors became thorough masters of the art of walking in hurricanes – an accomplishment comparable to skating or skiing. Ensconced in the lee of a substantial break-wind, one could leisurely observe the unnatural appearance of others walking about, apparently in imminent peril of falling on their faces.

Experiments were tried in the steady winds; firmly planting the feet on the ground, keeping the body rigid and leaning over on the invisible support. This "lying on the wind," at equilibrium, was a unique experience. As a rule the velocity remained uniform; when it fluctuated in a series of gusts, all our experience was likely to fail, for no sooner had the correct angle for the maximum velocity been assumed than a lull intervened – with the obvious result.

Before the art of "hurricane walking" was learnt, and in the primitive days of ice nails and finnesko, progression in high winds degenerated into crawling on hands and knees. Many of the more conservative persisted in this method, and, as a compensation, became the first exponents of the popular art of "board-sliding." A small piece of board, a wide ice flat and a hurricane were the three essentials for this new sport.

Wind alone would not have been so bad; drift snow accompanied it in overwhelming amount. In the autumn overcast weather ... heavy falls of snow prevailed, with the result that the air for several months was seldom free from drift. Indeed, during that time, there were not many days when objects a hundred yards away could be seen distinctly. Whatever else happened, the wind never abated, and so, even when the snow had ceased falling and the sky was clear, the drift continued until all the loose accumulations on the hinterland, for hundreds of miles back, had been swept out to sea. Day after day, deluges of drift streamed past the Hut, at times so dense as to obscure objects three feet away, until it seemed as if the atmosphere were almost solid snow.

Luckily, our field experience embraced only milder versions of these conditions, for we were there only in the austral summer. Even so, we often cursed the wind, in spite of the fact that the scouring winds of the winter, as described in the last paragraph of the extract above are what keep the ice patches clear of snow and make our work there possible in the first place. While we were grateful for that aspect of the weather conditions, it did not keep us from suffering from wind chill.

Mawson understood that one chills faster in the wind than in still air. Paul Siple named this phenomenon the Wind-Chill Factor, and quantified it. Essentially, your body will chill at a certain rate at a given low temperature in still air. The lower the temperature, the faster you will chill. If the wind is blowing, however, your body will chill faster than it would in still air, and the effect is that of standing in still air at a lower temperature. The extra degree of chilling induced by the wind is the wind-chill factor. We were operating in air temperatures of -10 to $-25\,°C$, but the areas of exposed ice we were investigating had ice exposed for a reason: high winds kept them swept free of the snow that mantles most of the continent. High winds are a physical presence in Antarctica, plucking at your clothing, snapping the tent walls in and out, and searing any exposed part of your body as the wind-chill produces its effect. During our first year, I would sit in our tent in the mornings listening to the wind flapping the canvas and the sibilant whisper of ice crystals saltating along the surface, streaming by the lower parts of the tent in a neverending, mindless rush to get downwind. I would mentally cringe in anticipation of facing the outside world, heard so menacingly through the tent wall. Finally, I would marshal my will power and emerge for the day's work. Part of the adaptation process was learning that we could even *survive* in this place.

Certain bodily functions were the greatest challenge. Eating was always a great pleasure, of course, and except for lunches on the trail, was conducted comfortably inside the tent with the stoves on for warmth, as well as for preparing food. This pleasant activity, however, inevitably results in other bodily functions that are accomplished outside, at some distance from camp, in the lee of one's snowmobile, squatting over a hole in the snow with all one's lower clothing around the ankles, usually in a high wind. This is a procedure that anyone can carry out once, but I maintain that living with the certain knowledge that this will happen again, every day for weeks into the future, does tend to build character. Robert Fudali, a colleague from the Smithsonian Institution, has observed that there are two extremes in experience that tend to stick in the memories of antarctic veterans – pleasurable ones and character builders. Almost always, the character builders make the best stories.

Every season, we all had to spend two days undergoing survival training before we were allowed to go to the field on our own. We learned and relearned techniques of rappelling, techniques for climbing a rope, a couple of simple knots, crevasse recognition and rescue, the use of crampons and ice axes, how to rope up to travel in crevassed areas, and how to construct a snow shelter and sleep in it. Gary Ball, whom I met during my first survival training experience, was a New Zealand (otherwise known affectionately as a Kiwi) instructor in the McMurdo survival school. He later hired out as guide for a number of years with German field parties working in North Victoria Land, and I heard from him the following story about the fury of antarctic winds.

Gary had taken a field party ashore from the German research vessel, the *Polarstern.* A sudden windstorm blew up and they found themselves some kilometers inland in a snow-free valley as a howling katabatic attacked them with rapidly increasing strength. They erected their tents and prepared to wait until the storm had blown itself out.

The *Polarstern,* meanwhile, was experiencing difficulties. The ship was in relatively shallow water near the shore, and it suddenly became too dangerous to run for the open sea. The captain maneuvered the ship into the lee of a large rock, where there was partial shelter from the wind and the surging sea. The wind increased in velocity. The ship's anemometer registered winds of 200 km/h before it was blown away. The captain was at the wheel for 48 hours with his ship under power in order to maintain its position behind the rock. During the storm, a sailor who ventured on deck was blown overboard, landing on an ice floe and breaking both thigh bones in the fall. Using heroic measures, the crew was able to rescue him before the ice carried him away.

At the same time, Gary's party was in a desperate situation. Gary and his field partner had been using a pyramid-shaped Scott tent, and this shredded. They wrapped the larger pieces around themselves and crouched there for two days, hoping to survive. The other two members of the party had used a lower-profile tent with aluminum supporting rods. When the aluminum rods bent flat in the wind, these two spent two days in a flattened-out tent, lying in their sleeping bags. The one on the upwind side came out of it with large bruises all over his body from cobbles that came bouncing down the valley during gusts. They all survived this experience.

DARWIN CAMP

During the third field season, we had a chance to search not only in the Allan Hills region but also farther afield, and much farther south around the "headwaters" of the Darwin Glacier. The Darwin is a relatively small feature, as glaciers go, but it flows parallel to one of Antarctica's larger glaciers, the Byrd. The tremendous Byrd Glacier and the much smaller Darwin Glacier next to it have been described by Terry Hughes of the University of Maine, as possibly analogous to a boiler with a small water-level gauge on the side: just as the gauge on the side of the boiler reflects changes in water level in the boiler, so does the Darwin Glacier respond to changes in the Byrd. One therefore might be able to study changes in the Byrd by observing the Darwin. So for this reason, among others, the Division of Polar Programs had built a remote base camp along the northern side of the Darwin Glacier. This camp had jamesway huts instead of tents. These structures come in sections that can be put together to form a shelter of any desired length. Each section consists of wooden arches with canvas stretched over the top. Plywood floors and end walls are installed. There are no windows except in the end walls, so illumination must come from electric lights. The camp consisted of a number of these jamesway huts, including a small one enclosing an electric generator and larger ones containing indoor plumbing supplied by melted snow, a mess hall, darkened sleeping rooms and an area for working on maps and field notes. It also had a helo pad with three helos available.

The interesting feature to us was the presence of a number of large areas of exposed ice up on the plateau that were close enough to be easily accessible by helicopter. They were associated with such rocky formations as Butcher Ridge, the Finger Ridges, Turnstile Ridge, Warren and Boomerang Ranges, Bates Nunatak and Lone Wolf Nunatak. The density of meteorites on these ice patches was a little disappointing, although we found enough to justify another visit in the future. Also, we could see extensive moraine fields up at the edge of the plateau, between the headwaters of the Darwin and Byrd Glaciers. I still did not understand the lack of success during our first field season in the moraines at the end of the Upper Wright Glacier, and I thought we should spend at least a little time searching these new moraine fields.

Until this time we had been finding meteorites on ice patches where there were no terrestrial rocks present. Now we proposed to search for them in "rock city," where there was very little ice visible because the surface was mantled with terrestrial rocks. Ursula Marvin of the Smithsonian Astrophysical Observatory, stood in the middle of this rocky wasteland near the headwaters of the Darwin, wondering how to find a needle in a haystack. She decided that the first thing to do would be to become familiar with the range of terrestrial rocks present. That way she would know if a given rock was unusual and might be a meteorite. To start this process, she picked up her first rock. It was a meteorite! Does this mean that there are other meteorites in this moraine? Maybe not, but I would guess that it does. Finding meteorites in a moraine is hard work, but if we are conscientious we will eventually return and search that moraine, on our hands and knees, if necessary. But, hey, it's not my problem. Since I am now retired, I am happy to leave this chore to others.

Now and then something happens that makes one wonder about retiring sooner, rather than later. On December 7, 1978, we were about a week into the third field season. Shiraishi and I had arranged for a reconnaissance flight to Butcher Ridge and the Finger Ridges, but had found nothing. When it came time to return we realized that a low-lying cloudbank had crept up on us and now lay between us and Darwin Camp. Lt. Frank Staebler, the pilot, had been keeping an eye on the clouds, calculating that he could skirt them and get us back safely. Trying to skirt them, however, we found ourselves cruising out over the 25-km wide headwaters of the Byrd Glacier and traveling at right angles to the direction we wished to go. We couldn't turn the corner. Then we tried going over them, but they appeared to form a complete blanket over the terrain as far as we could see. Finally, Frank spotted a hole in the blanket and plunged down through it, setting us down on a flat-topped mountain called Harvey Peak, without enough fuel for the return to Darwin Camp, should that even have been possible.

The top of Harvey Peak was covered with 20-cm to 1-m sized boulders, with very few spots where Frank could have landed safely. He had found one such spot, and it was located only about 25 m from another similar spot, where a relief helicopter would be able to land. Frank had set down a little before he really had to, and now had enough fuel to be able to fire up the engine now and then to keep the battery fully charged. This was important, because we did not want to lose radio contact with Darwin Camp. We let them know the situation and they promised to send out a relief mission when the clouds had cleared, but in any case not before the following morning. It was easy to establish that there were no meteorites on the several acres that formed the surface at the top of this peak, and also that it was surrounded on all sides by vertical cliffs. We spent a little time push-ing boulders off the edge, listening to the satisfying crashes as they were smashed to pieces 100 m below.

All the helicopters had emergency packs and each of us also had our survival bags, with bearpaw mittens, extra socks and so on. We broke out a couple of small, lightweight tents and flimsy sleeping bags that give an extra layer of insulation around the many layers of antarctic clothing. We unpacked some food and heated it over the emergency stove. There were plates, cups and eating utensils. Life looked a little less grim.

At 6 p.m. it started snowing, so we all crawled fully dressed into our mini-tents and into our sleeping bags.

I have always been an accomplished sleeper. I slept until 7:30 a.m., when the distant beat of a helicopter rotor could be heard. The clouds had crept away during the night and a thin layer of fresh snow covered everything. We listened for quite a while, searching the sky as the helicopter seemed repeatedly to approach and then recede into the distance. Finally, walking to the cliff edge we looked down and saw them below us. They had been circling the mountain, look-ing for us around the base of the edifice, while we were up on top where we thought we could be seen more easily. A short message on the hand-held radio brought them up to our level in a hurry and they landed at the spot we indicated.

The pilot of our relief helicopter was Lt. Cdr. "Wu" Ferrel, who gave the appearance of having kept himself in superb shape by lifting weights – really heavy weights. Wu seemed to find it difficult to pass through a doorway without stopping, grabbing the doorframe over-head and doing five or six chinups. He would need this ability today, because the relief helo had a 250 kg drum of fuel in the cabin and the two helicopters were separated by 25 m of difficult walking over boulders with a range of sizes. Rolling the drum was out of the ques-tion – it would have to be carried. Wu took one end of the drum alone and two crewmen picked up the other end. Gasping with the effort, the three of them carried it over the boulders and set it down next to

FIGURE 3.2 The 160 kg Derrick Peak iron meteorite. This is the largest representative of a shower of iron meteorites found on the slopes of Mt. Derrick. It is not associated with meteorites of any other type. The initial discovery of fragments of this shower was made by members of a New Zealand field party from the University of Waikato, led by Michael Selby. (Photo by W. Cassidy.)

our helicopter. We refueled; we could go. We went. I stopped thinking about early retirement.

On the following day, a team of New Zealand geologists under Professor Michael Selby from the University of Waikato made a major find in an apparent ancient shower of iron meteorites at Derrick Peak, within sight of Darwin camp. They radioed us to come over, and we were glad of the opportunity. They had found six irons scattered over the snow-free slopes of the mountain, and, searching jointly with them, we found six more which they very generously agreed to let us keep. One was a magnificent 160 kg iron meteorite (Figure 3.2) covered with what appeared to be piezoglyphs (sometimes called pressure marks or thumbprints).

Piezoglyphs are produced by turbulent eddies in the scorchingly hot incandescent gas passing the surface of the meteorite during its entry into the atmosphere. Strangely, the surface of the meteorite was also punctuated by protruding mineral grains that were later identified as phosphides. It is hard to see how these grains could have preferentially survived the ablation process that eroded the surface of the meteorite during entry, so this added a little mystery to the find. The mode of occurrence of iron meteorites in Antarctica is a little mysterious anyway, and will be discussed again in Chapter 9.

RETROSPECTIVE

During the first few field seasons, we had a lot of success in recovering meteorites. During this time, we were in the field with Japanese colleagues who had a lot of experience in Antarctica. From them I learned how to survive on the ice plateau of East Antarctica and was able to measure my abilities and limitations under severe conditions. It was also a period during which personnel from the Division of Polar Programs had to form their ideas about us and the ANSMET project. These ideas, in turn, would influence the allocation of resources devoted to our field efforts. It was important that they should value our project highly because, in those days, the antarctic scientific community was much smaller and quite difficult to break into. In a real sense, we represented intruders who were after a piece of the pie, and we probably were as successful as could be expected under those circumstances.

Luckily, Mort Turner was convinced early on of the profound effect the ANSMET project was having on the planetary science community. At that time, however, this conviction was not generally held within the Division of Polar Programs, which seldom had dealings with planetary scientists. This has improved with time, but we have always had difficulty in obtaining the logistics resources that would have allowed us to be as successful as we could have been. I believe I can say this in an unbiased manner, in spite of the fact that every principal investigator worth his salt feels that his project is just a little bit more deserving than anyone else's. During the first two years of the project, in spite of repeated requests, we were not provided with snowmobiles and had to carry out all our searches on foot, walking between our camp and the areas we wished to search. Our two Japanese colleagues could not understand this, and for the third field season they sent two Japanese snowmobiles to McMurdo for their own use on the ANSMET project. Once again, they had helped us out: the Division of Polar Programs could not allow its two participating scientists to become mere hitchhikers across the frozen wastes, so we also got snowmobiles.

By the end of the first three years for ANSMET in Antarctica, our pattern of operations had been established. We would do one of two things: either we would spend part of the season collecting meteorites in known areas of concentration and another part in exploration for new sites, or we would split the field party in two, with one part doing reconnaissance and the other part doing serious collecting. This would ensure that we always had a successful current season, and we always had new sites in the bank where we knew we could go for a successful future season.

In the first three years during which US scientists were actively searching for meteorites in Antarctica, we continued our cooperative program with Japanese scientists. The joint team operated out of McMurdo Station, principally at sites within helicopter range of the US base, and during those three years we recovered 658 meteorite specimens. The new sites were located across the continent from the Yamato Mountains, and this provided positive confirmation of the hypothesis that the Japanese discovery site was not unique to the continent. It also suggested that meteorites might be found at many other localities in Antarctica. Since the end of the 1978–79 field season, the Japanese and US groups have operated independently again, pursuing their own programs and finding new meteorite concentration sites with great regularity. The entire experience has been remarkable.

4 The beat goes on: later years of the ANSMET program

FIELD LOGISTICS

If field work is to be carried out within 100 nautical miles (= 185 km) of McMurdo Station the preferred mode of travel is by helicopter, but we had begun prospecting for meteorites at sites that were out of helicopter range. For a while, it was sufficient to be put in at Allan Hills by helicopter and travel from there by snowmobile, towing everything on sledges. Our snowmobiles were geared-down machines made by Bombardier Corp. of Canada and were designed for heavy pulling. We found we could tow three fully loaded nansen sledges about as easily as one, so our cargo transport capacity gave us self-sufficiency for long oversnow traverses and long stays at remote sites (Figure 4.1). In this way, we were able to work effectively at the Reckling Peak, Elephant Moraine and Allan Hills Far Western icefields (see Frontispiece). But existing satellite photos gave us the ability to identify ice patches in all parts of the continent, and there were more-distant places that we aspired to visit. Camps at these sites are often referred to as deep-field camps.

Longer lifts are carried out by LC-130s, which actually can reach any part of the antarctic continent. For extreme distances, there is a trade-off between cargo weight and distance flown, but that limitation has not yet affected our field operations. The LC-130s (Figure 4.2) have been fitted with aluminum skis. These are very large, commensurate with the size of the airplane, and have been coated with Teflon®.

An open-field landing – that is, a landing at a site where there is no prepared runway – is a ticklish operation. Things can go wrong that will destroy a plane, or damage it so severely that it cannot take off again. The surface must not be too hard, too uneven, or crevassed.

FIGURE 4.1 1988 traverse by snowmobile and sledge from Beardmore South to our deep-field camp at the Lewis Cliff Ice Tongue. Due to the lack of wind at the Beardmore South site the snow is unusually smooth, with no sastrugi for the first 30 km. (Photo by W. Cassidy.)

As a first step, the pilot approaches the area of the desired put-in point and looks for a smooth surface. He flies low over his projected landing field and looks for bumps, ice patches, and any evidence of crevassing, which can be indicated by bands where the snow appears to have settled a little below the surrounding surface. If none of these characteristics can be seen, he then does a "ski-drag." This consists of flying along the surface dragging the skis, with almost no weight pressing on the snow, trying to feel the texture of the surface. He is doing this with an aircraft that with its cargo might weigh as much as 53 metric tons. At the end of the ski-drag he lifts up to about 20 m and flies back along the track he has made in the snow. If it looks like a dashed line (— — —), or a dotted line (. . . . . . .), the surface is too uneven and he tries somewhere else. Subjective judgment is involved; if it felt too hard or too rough, he tries somewhere else. If it felt smooth and the track looks pretty continuous, he must make

FIGURE 4.2 An LC-130 cargo plane getting airborne on the relatively easy surface provided by the smooth snow at the Beardmore South site. This site was used as a remote base camp during the 1984–85 and 1985–86 field seasons. The surface here is well known to LC-130 pilots, who commonly use it to practice open-field landings. We used it as a put-in site for the 1988–89 field season because a survival shelter had been left behind, and it had not yet become buried by the annual accumulations of new snow. (Photo by R. Korotev.)

a decision to commit to a landing. This is a real commitment, because, once his airspeed drops below minimum flight speed, he cannot take off again; whatever is ahead will happen. These landings are often described as "controlled crash landings." They are pretty much the supreme test of a pilot's skill (and of the passengers' bladder control).

Once on the ground safely, the plane turns around, making a loop at the end of its track; the back end of the plane opens and the crew pushes the palletized cargo out while the plane is still taxiing. This is called "freight-training it out." Pallets with 55-gallon (approx. 200 liter) drums of fuel, boxes of food, equipment, snowmobiles and sledges slide down along rollers and out onto the snow, almost as if they are hooked together like the cars of a train. Strangely, one of us must sign a bill of lading to verify that everything has been delivered, and keep multiple copies for our files. Only the accountants know why.

We clamber out, the plane gets back in its track and retraces its landing path, returns and lengthens the track, and taxis back and forth a few times to pack it down. Essentially, it makes its own runway. Then, off in the distance, it begins its takeoff run, sweeps past us onlookers and zooms into the air. One more pass in the air to waggle its wings at us – we wave, and it is gone. Suddenly, it is very, very quiet.

Once we have arrived, it has not been unusual to put 1000 km or more of travel on each snowmobile during a field season, with a combination of travel from the LC-130 put-in site to the ice patch, systematic searching while at the site and travel from site to site. Frequently we can get parachute delivery of snowmobile fuel at points along the route (Figure 4.3), and this service helps with our logistics problems. It also helps with our morale, because mail accumulated at McMurdo is air-dropped at these times.

Snowmobiling in Antarctica is seldom a smooth ride. Wind-blown ice crystals can accumulate in dunelike structures called sastrugi (Figure 4.4).

These dunes tend to be elongated parallel to the direction of the prevailing winds and extend in length from a few meters to hundreds of meters. They can be 50 cm to 1–2 m in height. Their hardness may be due to efficient packing of ice grains, or to sintering, both of which can impart a degree of structural strength to a previously unconsolidated mass of crystals.

Sastrugi can occur in endless parallel ranks, and while they make for a rough ride, their presence in most places can be an aid to navigation. Once your direction of travel has been established, you can maintain a straight course through trackless wastes just by maintaining the same angle with reference to the sastrugi. You can also have a general idea of your direction because they tend to run south to north, just because the prevailing winds that formed them blow approximately from south to north.

FIGURE 4.3 Two pallets, each containing four 55 gallon (approx. 200 liter) drums of gas is air-dropped at Reckling Moraine. The pallet is suspended below a high-speed parachute and strikes the surface at about 90 km/h. Other available parachutes would give it a softer landing, but could reinflate once the load is on the ground. If that happened, and the pallet had landed on ice, a strong wind might drag the load a long distance before we could stop it. The drums are cushioned from the shock of landing by a layer of corrugated cardboard about 25 cm thick, with the corrugations oriented vertically. (Photo by W. Cassidy.)

Traveling due west or due east, the trip is an endless series of ups and downs as we cross sastrugi at close to right angles. Actually, this is a good course to take if one is towing sledges, because sledging has its special challenge relative to sastrugi. A sledge trails its snowmobile at the end of a 6- or 7-m rope; a second sledge trails the first at a similar distance. At right angles to large sastrugi, one has to deal only with the increased strain as the sledge is pulled up one side of the dune and the sudden relaxation as it slides down the other side. To keep tension on the rope and to keep the sledge traveling in the right direction, one must accelerate at the correct time. This is seldom done perfectly and, when given a slack rope, the sledge will always veer off to one side or the other. If allowed to veer too far before the slack is taken up, the

FIGURE 4.4 Sastrugi, or snow dunes, ensure that snowmobiling and sledging in Antarctica never become boring. (Photo by W. Cassidy.)

next tug on the rope can whip the sledge back into line so quickly that it turns over. When that happens everyone has to stop to help in righting the sledge.

Crossing sastrugi at less than 90° is more difficult because ascending at an angle causes a certain amount of sideslip and redirection of the sledge that must be corrected on the downslope with just the right amount of tension on the rope. Some people can become extremely adept at this, but I never became more than mediocre.

The real test can come when you must descend a long hillside. Tension must be maintained on the rope in order to keep the sledges from veering away. Since the sledges would slide down the slope without your help, you must constantly accelerate, and so do the sledges. At the bottom of a long hill the train can be running quite fast and nansen sledges will flip over more readily, for all the usual causes. On occasion during a descent you will watch while one of your sledges passes you, and you know then that all is lost.

THE CONCENTRATION PROCESS

After the first few years of the ANSMET project, we better understood some of the reasons why meteorites could be concentrated more in Antarctica than in the rest of the world. Partly, this resulted from the first, early measurements of the terrestrial ages of some of the Allan Hills meteorites by Lou Rancitelli and his colleagues at Battelle Institute Northwest in Richland, Washington.

How can one estimate how long a meteorite has been on the earth, i.e., its terrestrial age? When meteorites fall on the earth, they arrive with an acquired cargo of ^{26}Al, which is an unstable isotope of aluminum that has been produced in the meteorites by cosmic ray bombardment during their long sojourn in space. Most meteorites have been in space, orbiting the sun as small bodies, for a long enough period that an equilibrium has been reached between cosmic-ray production of ^{26}Al and decay of the same isotope. By having measured the ^{26}Al concentrations of hundreds of freshly fallen meteorites of all types, we know what this equilibrium concentration is for a given type of meteorite.

When the meteorite reaches the earth's surface it is shielded from cosmic rays by the earth's atmosphere and the earth's magnetic field, so production of ^{26}Al stops, but loss of ^{26}Al by radioactive decay continues, at a known rate. If a meteorite has been on the surface of the earth for a long time, its ^{26}Al content will be depleted below the equilibrium level that it probably had when it arrived from space, and the degree of such depletion is a measure of its terrestrial age.

Most meteorites weather away within several hundred to several thousand years on earth, before they have lost much of their original ^{26}Al. Rancitelli and his colleagues, however, were finding that many antarctic specimens showed significant depletions in ^{26}Al, signifying that they had been on the earth's surface for great lengths of time. We have to conclude that antarctic meteorites have not weathered away at such a great rate as they do in the rest of the world; they apparently are quite comfortable in the deep-freeze conditions in Antarctica, weathering only very, very slowly. The measured terrestrial ages of antarctic meteorites range between about 10,000 years and two million years (!), although meteorites with terrestrial ages of one to two million years are very rare. Because they can survive for such great periods on the ice of Antarctica their numbers can accumulate through time, leading to possible concentrations on the surface. But this was not the complete answer to our questions about the concentration process.

A collaboration that considerably helped my understanding of the concentration mechanism of meteorites on the ice sheet resulted from a series of conversations I had in 1982 with Ian Whillans, of the Byrd Polar Research Institute at Ohio State University. Ian died in 2000, much too early, of an inoperable brain tumor. Ian was a glaciologist, and had a real understanding of the physics and dynamics of continent-size masses of ice. Early on, he had recognized the potential value of the antarctic meteorite occurrence and had started thinking about the relationship between the meteorites and the ice sheet. We discussed flow in the ice sheet, and from this discussion developed a model for the pathway that a meteorite follows from the time it falls in Antarctica until it leaves the continent in an iceberg, eventually to fall to the ocean floor as the ice melts.

Suppose the meteorite falls near the center of the continent at, say, 3500 m elevation. It initially buries itself in the snow. During succeeding seasons more snow falls, burying it more deeply. When it is about 30 m deep the snow begins to undergo pressure recrystallization and to form ice, squeezing out a lot of the air between the grains. The meteorite is now embedded in ice.

As time passes and more snow falls, the meteorite follows an ever deeper trajectory and moves outward with the ice toward the edge of the continent. In this way, most meteorites that fall onto the ice sheet eventually are buried at sea. What a terrible waste of scientifically interesting material! The exceptions, which are the ones we find, are associated with patches of ice that cannot reach the sea because they are trapped behind a barrier.

Ice has nowhere to go if its path has led it to a barrier, such as a mountain range, that it cannot flow over or around. If there are sufficiently high winds that blow pretty constantly at this site, the overlying snow layer is stripped away and the ice is exposed. Now two things happen: the constant wind blows ice crystals over the surface, producing a sandblasting effect that wears the ice away and, in the summer, solar heating causes vaporization at the ice surface, which also wears it away. The wearing-away process is called ablation. So even though ice trapped behind a barrier cannot flow away, it can leave by ablation.

It is important now to understand that ice acts like a sluggish hydraulic system; as surface ice leaves by ablation, upstream ice moves from the interior of the continent into the stagnant zone to restore the original surface elevation. So essentially, in a zone of stagnant flow, ice can leave by passing through the ice–air interface, and, in the absence of serious climate change that would upset this equilibrium, the supply of new ice arriving at this interface always is just the amount that has left by ablation. The significance of this to meteoritics is that if an absolute barrier to ice flow exists for a very long period of time, it is certain that the very occasional meteorite will be brought up by the ice and will be left behind, stranded on the ice surface. For this reason, we refer to these patches of ice as *meteorite stranding surfaces.*

Now, because meteorites weather away so very slowly in Antarctica, it is easy to see how a concentration of meteorites can be created on a meteorite stranding surface. The essential ingredients are stagnant ice flow, high winds and very large chunks of time. Because the stranding surface must have been there for a great length of time, it may have accumulated meteorites that fell directly onto it, in addition to those carried there by the ice. Combining my experience of the field occurrence of meteorites in Antarctica with Ian's perceptions of ice sheet dynamics, we had a paper published in *Science* called "Catch a Falling Star: Meteorites and Old Ice." But even this was not the final answer.

Over the years, we have found that complete stagnation of ice flow is not absolutely necessary to form meteorite concentrations. Most of the occurrences we have discovered, in fact, are not located behind perfect barriers to flow. In many field situations the ice of the stranding surface seems to be descending a step and flowing away. We believe this downward step is caused by ice actually overwhelming a subsurface barrier that is not high enough to completely restrict flow. In these cases, ice is flowing over the barrier and still a concentration of meteorites is found on the surface. It seems necessary to assume that the barrier slows the ice down and high winds cause ablation fast enough to allow a concentration of meteorites to accumulate at these places. One such place is the Elephant Moraine stranding surface.

HOW DO YOU CHOOSE FIELD PERSONNEL?

One way is to rely on chance encounters. For example, John Annexstad served for a number of field seasons. I had met John after our first field season, when he called me on the telephone from the NASA/Johnson Space Center (JSC) in Houston, where he was an Associate Curator of Lunar Samples. He had invited me to visit, in order to assess their curatorial capabilities. He said very frankly that they hoped to convince me that the antarctic meteorites would be better processed there than anywhere else, using methods similar to those used for the lunar samples. I had found John's frankness refreshing, and ended up convinced. As it turned out, of course, I had no influence on that decision, but it had been nice to meet the group at the Lunar Receiving Lab.

FIGURE 4.5 Aerial photograph of the Elephant Moraine ice patch. White streaks and patches are snow and grey tones are exposed ice surfaces. The darkness of the grey reflects the density of terrestrial rocks scattered over the ice surface. The arrangement of surface rocks gives this moraine the appearance of an elephant with an abnormally long trunk; hence its name. The "trunk" is at least 7 km in length. There is no rock outcrop at Elephant Moraine. The moraine consists of rocks carried up from bedrock underlying the ice, and the elephant-like outline, while remarkable, is fortuitous. (USGS aerial photo.)

John had had earlier antarctic experience, with one of the US International Geophysical Year expeditions and also with the Australians, so he was a natural choice as a field partner. He first joined us during the 1978–79 season, and, after seeing some of the terrain we were skirting, made an important recommendation that we always go to the field with a crevasse expert. This made a lot of sense, and we have done this every year since John made the suggestion. Our first crevasse expert was Lee Benda, a mountaineer from the state of Washington. He and John shared a tent during the 1979–80 season, while I shared a tent with Lou Rancitelli, of Batelle Institute. This was our entire field party at that stage of the field season.

By that time I had pretty much established my policy on field personnel: we would always take along a crevasse expert; after that, first priority would be given to senior people who were actively involved in meteorite research; second priority would be given to graduate students in meteoritics and allied fields; then if there were still openings, I would consider applicants with special experience such as first aid, communications, electronics, snowmobile maintenance and so on. I had decided to include non-US scientists because science is an international endeavor. I also felt that we should try to have someone with previous antarctic experience in each two-person tent.

This policy has always seemed to work very well. Strangely, many of those participants who were most successful were laboratory investigators with no previous field experience. I attribute their success to two factors: from their background in laboratory research, they understood better than anyone else how important it is to avoid contamination of samples, and because they did not have preconceived ideas about how field work should be conducted, they readily adapted to our established procedures. In addition, however, there was an intangible factor involving intellectual satisfaction. This was best expressed by Ludolf Schultz, of the Max Planck Institute for Chemistry, in Mainz, Germany. Ludolf once said, "Bill, before I came to the Field in Antarctica a Meteorite was a little Bit of Powder that someone sent me in a Vial. Now I have seen Meteorites as they occur in Nature, and I have a great new Appreciation of what they are and of their Importance to Science!" (Ludolf, being German, always capitalizes his nouns, even when speaking.)

Now that I think of it, Ludolf had two other assets that made him an outstanding field person in Antarctica: excellent blood circulation in the extremities and a very good sense of humor (excellent blood circulation in the brain). Now and then in the field, we all have to bare our hands for a brief period, and not only during the regular crouching-by-the-snowmobile-over-a-hole-in-the-snow routine. Many times, the tape we use to seal the sample packages is firmly stuck to the roll and we have to remove our gloves to get a fingernail under the edge. Also, most people have to remove their mittens to write field notes, and whenever you remove your gloves or mittens, your hands immediately start to freeze. Thanks to excellent circulation, Ludolf was demonstrably slower to freeze his fingers and faster to recover after putting his gloves back on.

In the sense of humor department, I once told Ludolf a well-known story about how to get through an Antarctic winter with only two sets of underwear and no way to wash them: wear one pair for two weeks; then change to the second pair and wear them for four weeks. By that time the first pair seems clean, so change to that pair for the next four weeks, and so on . . . Ludolf came right back with a German version of this story: in a German winterover Station, everyone wears the same Underwear for four Weeks; then one Morning, a Clerk comes out with a Clipboard and says, "We will all now change Underwear. Willi will change with Dieter; Hans will change with Rudy; . . ."

Having a sense of humor is very important under trying circumstances, and circumstances can be trying in the field in Antarctica. John, Lee, Lou and I had just traversed 90 km from Reckling Moraine to make camp near the end of the elephant's trunk at our new site. We were tired, and glad only of the chance for dinner and a good night's rest. For the following day, all of us were curious to get up to the eye of the elephant, where we believed there was a rock outcrop, to see what kind of rock it was. In our tent, over dinner, Lou and I fantasized about racing Lee and John up to the outcrop the next day, in order to be the first to raise a marker flag on the exact top of the rock. We decided to recreate in a minor way the events at the South Pole, when Amundsen outraced Scott to be first to attain 90° S. Amundsen left two letters for Scott: one addressed to the King of Norway, announcing his attainment of the South Pole and the other addressed to Robert Falcon Scott, asking him to deliver the first letter to the King of Norway. This would be proof certain that Scott had gotten to the pole second, in order to have found the letters left there earlier by Amundsen. Being only second to reach the pole must have been partly responsible for Scott's famous quotation, found in his fieldnotes, describing the South Pole: "Great God! this is an awful place . . ."

That evening in our tent, Lou and I concocted two letters: one addressed to Mort Turner, Program Manager for Geology in the Division of Polar Programs, announcing that we had been first to reach the Eye of the Elephant at Elephant Moraine, South Victoria Land, Antarctica and the second, addressed to John Annexstad, assuring him that being second was better than being third, and asking him to deliver the first letter to Mort Turner. We chuckled a lot over this bit of cleverness and suddenly we didn't seem as tired as we had before.

The next morning we got up an hour early and stole a march on John and Lee, who were usually out ahead of us. We drove up along the moraine on our way to the Eye and suddenly found a meteorite; then another and another. Well, you can't ignore a meteorite, so we were busy collecting them when John and Lee cruised by and went up the slope. They strolled over to the Eye and I never got to use the two letters. Until now.

RUNNING A FIELD PARTY IN ANTARCTICA

After the first three or four field seasons, during which I had selected field team participants from a pool of scientists and graduate students engaged in meteorite research, I felt quite satisfied with our performance. An exception to the general run of field participants was John Annexstad, who had already accumulated a much wider range of antarctic field experience than the rest of us. In addition, he came to us from an administrative position within NASA and he understood better than I how government agencies worked. I could only consider these qualities to be assets, and they were, for the most part. I did not understand at first that when you take aboard someone with John's qualifications you also get someone who is likely to disagree with your judgment, from time to time. I have often thought how hard it must have been for John to defer to some of my decisions, but a field team must have only one leader and, when all was said and done, I was the Principal Investigator.

John's sense of discomfort came to a head, eventually, and he decided to submit a proposal to augment the ANSMET program with a program of his own devising. If successful, his proposal would bring more money into the field effort, assuming that mine would continue at the same level. Potentially, this could have been a benefit to the overall effort. The objection I had to it was that the entire field operation would have to be reorganized into two quite different groups. There would be a reconnaissance group that John would direct and a field collection group of which I would remain the director. I made my objections known to John and to Mike Duke, John's boss at NASA, but Mike supported John's endeavor.

I experienced a distinct feeling of guilt over the sense of relief I felt to learn that John's proposal had not been funded, but I was successful in repressing it (the guilt, that is). Later I received a generous letter from Mike Duke. The main paragraph follows.

Dear Bill:
   Thank you for the copy of your proposed continuation and augmentation of your NSF program. The approach is quite different from the one we suggested to you. I am sorry that you did not accept the basic premise of separating reconnaissance and collecting efforts to make the most efficient use of resources and

people. This remains a difference of opinion between us, but we do not intend to pursue the program independently and will remain silent on the question of long term strategy and planning unless asked specifically by you or the NSF. We will continue to support the Antarctic meteorite program to the best of our ability.

In thinking about this experience I am convinced I had the right instincts. We use very intelligent people in the field teams, and they need change from time to time. Systematic collecting day after day can be sustained, but if done for week after week it becomes just another boring activity. Intelligent people need an escape from boredom, and reconnaissance trips can provide that. This was demonstrated to me forcefully during the 1988–89 field season. We had been searching systematically on the Lewis Cliff Ice Tongue for several weeks, and finding sparsely distributed, small, ordinary chondrites on the ice among a decent-sized population of small black rocks. The work had become deadly dull, because each black rock had to be examined closely to make sure it was not a meteorite, and it almost always was not. I decided we should take a day trip up to the MacAlpine Hills, at the headwaters of the Law Glacier, for a break. The snowmobile traverse was fun in itself, but when we arrived we could see a smallish patch of ice, partially snow-covered, with a lot of rocks on it, *and very few of them terrestrial.* We started a systematic search pattern, but there were too many meteorites, and team members started acting like sharks attending a feeding frenzy. They ran on foot from meteorite to meteorite, planting marker flags next to each new discovery while looking around and spotting two or three new ones for every one they were marking. They were shouting and laughing. This experience provided a real emotional release from the frustrations that had been building for more than a week before, and I realized they had needed this break sooner – I had waited too long. Once again, I felt a sense of relief that we were an all-purpose field party; not limited to routine collecting for the entire time that we were in the field.

ANTARCTIC METEORITE NAMES

When meteorites are found in the rest of the world, they are named after the nearest post office, because a place that has a post office is likely also to be indicated on a map. Antarctica, however, has many, many more meteorites than post offices, so we try to name the meteorites after a mappable, or already mapped, geographic feature. At a meteorite stranding surface, however, there are many more meteorites even than geographic features, so we name the stranding surface and give the meteorites from that stranding surface a temporary field number. In the arcane naming system used for the antarctic meteorites, the full name is never spelled out because the name is entered into a computer. Elephant then becomes EET (the first, third and last letters) plus a number, for example 79001. In this number, 79 is the December year of the field season during which the meteorite was collected and 001 indicates the sequence in which the meteorite was examined at JSC, not the sequence in which it was originally found. Thus, EET 79001 was the first Elephant Moraine specimen examined at JSC from the collection made during December 1979–January 1980. Some of the earlier stranding surfaces were given arbitrary three-letter designations, such as ALH for Allan Hills, and more recently, we have tried to use the first three letters of the name, for example LEW for Lewis Cliff. All this reflects the fact that a program that grows from a low level of expectation to reach an unpredicted degree of success may not have been preplanned in great detail, and often shows signs of improvisation along the way.

ANTARCTICA AS A MARS ANALOG

The Dry Valleys are probably the closest feature in Antarctica to a Mars analog. If the Dry Valleys are an analog to most of Mars, then the ice plateau of East Antarctica is the closest analog to the martian residual polar caps.

In 1984 a group of diplomats were thrust into this environment, in a remote field camp on the Bowden Névé, less than 600 km from South Pole. These were official delegates to a regularly scheduled meeting of the Scientific Committee on Antarctic Research (SCAR). All the Antarctic Treaty nations send representatives to these meetings, to discuss their stewardship of the continent. Traditionally, the usual meeting places are centers of civilization like Paris, London and Tokyo. But the conference that year was being hosted by the United States, and Division of Polar Programs had decided to convene it in the actual place that was being discussed – quite a radical idea! Many scientists in the U.S. program had been requesting a remote field camp near the Beardmore Glacier. The Division of Polar Programs had agreed to build one on the Bowden Névé, just west of the Beardmore, for the 1985 field season, so they built it a year early to house the SCAR conference.

Three principal investigators in the US program had been invited to visit in order to present short talks about their current antarctic research, and I was one of these. It was interesting to be able to walk around in the snow with people whose main business in the past had been conducted while going to parties at each other's embassies, or so I thought.

I have always wondered what *they* thought of this experience. I suspect many of them were not enthusiastic. At any rate, while I was visiting with an international group of diplomats, dumped down on what may have seemed to them to be – well, Mars – the rest of the field party was harvesting meteorites at the Allan Hills Far Western Icefield, and they had just picked up a 1.9309 kg meteorite for which, in 1996, the claim would be made that it was from Mars and that it might contain fossil remains of life on Mars. Can life exist on Mars, and, if so, is it worth living? Well, can life exist on a Mars analog and, if so, is *it* worth living? One could ask the delegates to the SCAR meeting, but must expect only a diplomatic reply.

In 1985 the Beardmore South Camp, as the remote camp on the Bowden Névé was called, came into use as a remote base camp for a number of field parties; among them were four of us from the ANSMET project. The location of this camp came originally from a suggestion by David Elliot, of the Byrd Polar Research Center at The

Ohio State University. As I have indicated earlier, David seemingly has been everywhere and, in addition, knows all there is to know about Antarctica. I think I have never been to a place where David had not been earlier. In these days, it seems to be impossible in Antarctica to "go where no man has gone before" because David Elliot already will have been there. David put his finger on the map where the Beardmore South Camp was later built, and said, "This is the best place I can think of in this entire region. I think we should put the camp here." So it was done.

The site of the Beardmore South Camp enjoys a splendid microclimate, at least during the summer. The scenery is gorgeous. The camp was located on a sort of windless, snowy plain with the Queen Alexandra Range to the east and the Queen Elizabeth Range far to the west. About 50 km to the southwest, one could see Mt. Achernar in the dim distance. We had no premonition, but this was to be the site of our next major find – the Lewis Cliff Ice Tongue, located just below Mt. Achernar. From our field camp near Mt. Achernar, which typically was windy and cold, we could close up our tents, hop on our snowmobiles and zip northward to Beardmore South Camp for a few meals cooked by others than ourselves, a shower and maybe a movie. About halfway along this route to the Beardmore South Camp, exactly at the point where we passed Mt. Sirius on our journey north, it was as if we had gone through an airlock: the wind died completely. It did not resume until we went back through the airlock at the same place on our way south again. We thought of these visits to the Beardmore South Camp as a weekend at the beach, complete with beach volleyball in the snow, out between the jamesway huts and the LC-130 taxiway.

HARDTIMES CAMP

At the Beardmore South Camp, I had submitted a list of sites I wanted to visit that were within helicopter range of the camp, and so could be run as day trips. I had also requested that our party be put in at Otway Massif and picked up later at the Dominion Range. Some of the other field parties had priority over us, so we spent a few days making day trips by snowmobile in the direction of Mt. Achernar, where the aerial photos had shown a lot of exposed ice. On December 10, 1985 we were enroute to Mt. Achernar the first time, keeping partway up the snow slopes at the base of a series of low, rounded hills stretching between Mt. Sirius on the north and Lewis Cliff to the south, when we saw a couple of black dots out on the Walcott Névé, to the east. Black dots where there should be only white snow have to be interesting, so we drove down to them, speculating that they would turn out to be 55-gallon (approx. 200 liter) fuel drums left there by some earlier expedition. If they still contained fuel, they could even be useful. Coming up on the site, we saw the dots resolve themselves into a plywood shack – a 4 × 2 × 1 m plywood box, really – that had a fuel drum outside it. A fuel line led into the box and electrical wires led out, so this had been an electrical generator shack. The structure had been nailed shut, but there had been wild winds and blowing snow. The area on one side of the shack was a wind scoop, kept clear by eddies in the gusting wind, but the shack itself was full of snow, filled grain after grain by snow crystals seeping in through cracks until the inside could hold no more.

Near the generator shack, scattered around in a somewhat regular arrangement, were poles sticking up out of the snow, each made from two 2 × 4 inch (c. 5 × 10 cm) pieces of lumber, nailed together, with stapled-on electrical cables snaking up them but cut off at the tops. The wood had been dessicated over many seasons (Antarctica is a desert, after all, with terribly dry air) and had lost almost all its strength. We found in one area that we were walking around on the barely covered tops of buried fuel drums (probably empty).

This tiny part of Antarctica had once been occupied by people: it had been someone's field camp. This had been A Place! A bamboo pole with tattered red ribbons tied to it that had once been part of a trail flag marked a cache of food, left behind when the inhabitants departed for the last time. Who had they been? Why had they been there? Over toward the edge of The Place we found a weatherbeaten sign, with barely legible lettering that told a seemingly plaintive story: the name of The Place had been "Hardtimes Camp."

Back at Beardmore South Camp, full of our story, we found two people who knew Hardtimes Camp well: David Elliot (of course) and Jim Collinson, a paleontologist/sedimentologist from Ohio State who deciphers ancient environments. Hardtimes Camp has a history.

In 1969 David Elliot was leading a group whose principal goal was to seek vertebrate fossils in the Beardmore Glacier area. Why this part of Antarctica? Two years earlier, Peter Barrett, a New Zealander who was a graduate student at Ohio State University, had found a piece of the skull of a labyrinthodont, a fresh-water amphibian that had lived on all the southern continents 200 million years ago. This find had supported the possibility that Antarctica had once been connected to the southern continents. The find had been made at Graphite Peak, on the eastern side of Beardmore Glacier. David had planned two remote base camps, located so that many exposures of rock of the same age as that of the labyrinthodont would be accessible by helicopter. The site, later known as Hardtimes Camp, was placed on soft snow, conveniently close to a recognizable landmark known as Coalsack Bluff, so that aircraft could find it easily and land on a hospitable surface.

This camp was the first of the two to be constructed by Navy Seabees. The second camp, closer to Graphite Peak, was never built because David found an ancient river bed buried in the strata at Coalsack Bluff, only a few km away from the first camp, that seemed to be littered with enough fossilized vertebrate remains to occupy all the expedition members for more than a complete field season. These fossils represented an assortment of different animals, leading to the firm conclusion, because of commonality of occurrence, that Antarctica had been connected to Africa, South America, India and Australia for at least 60 million years. Since all these continents now are separated by formidably wide expanses of ocean, the old ideas of continental drift and an ancient supercontinent, christened Gondwanaland, came surging back with renewed vitality.

All this had happened at Hardtimes Camp. These were major discoveries, and I was curious why they had not named it Goodtimes Camp, instead. The story, from Jim Collinson, was that they had only one movie in the camp, so they watched the same film every evening for the entire field season. It was an old movie starring Henry Fonda, called *Welcome to Hard Times* . . . and it became forever linked in their minds with the camp at Coalsack Bluff. Actually, it was not an inappropriate name as far as weather was concerned. Jim recounted that when he worked at the fossil preparation tables in the jamesway that housed the field laboratory, snow would be always blowing across the floor and he could accurately gauge the severity of the weather outside by noting how fast the snow was swirling around his shoes. Hardtimes *and* cold times!

DISCOVERY OF A STRANDING SURFACE AT THE LEWIS CLIFF ICE TONGUE

The day after we rediscovered Hardtimes Camp, we retraced our tracks to Coalsack Bluff and continued onward to a tongue of exposed ice below Mt. Achernar. This ice runs parallel to a line of cliffs called Lewis Cliff, which rises out of the ice from the south and soars ever higher until it culminates in the crest of Mt. Achernar. There were four of us: Peter Englert, a nuclear activation chemist from San Jose University in California; Twyla Thomas, a technician from the Division of Meteorites at the US National Museum; Carl Thompson, a New Zealand crevasse expert; and me. We were delighted to find 12–15 meteorites almost immediately, and stopped to collect five of them, marking the others for later retrieval. At the beginning of the field season when you start finding meteorites, most people express a degree of enthusiasm that stops just short of turning somersaults. Twyla's reaction was quite different; it was absolute awe, which didn't seem to diminish for days. Twyla was a delightful field person. Every time we found a new meteorite, she would circle around it, exclaiming, "Ooh! Gee! Wow!," and eventually dared to approach it closely for a look at the details. She became very proficient at spotting meteorites, but this initial reaction was reward enough for including her as a member of the field party.

Before we could do much to exploit the site at Lewis Cliff, which we recognized as a major new find, we got the opportunity I had asked for to be put in at Otway Massif, and were occupied for the next 15 days with a reconnaissance traverse from Otway to the Dominion Range, visiting and searching a large number of ice surfaces along the way as we skirted the "headwaters" of the Mill Glacier, and finding a significant couple of dozen meteorites. We kept thinking, however, about the unfinished business at Lewis Cliff.

Back at Beardmore South Camp, we made a beeline to the Lewis Cliff Ice Tongue, putting our camp at the Hardtimes site. We supplemented our food supplies from their food cache, which consisted mainly of a couple of one-gallon cans of grated cheddar cheese and about a dozen one gallon cans of mandarin oranges. Carl Thompson ("Thomo") liked the mandarin oranges tremendously and ate so many of them he came down with diarrhea – definitely a bad idea in Antarctica.

The Lewis Cliff Ice Tongue eventually was to yield close to 2000 meteorite fragments, over a number of field seasons. While we were finding success at the Lewis Cliff Ice Tongue, however, the rest of the 1985 field party was having rough going. We had four people out at Allan Hills. For the six weeks they spent in the field, bad weather kept them in their tents fully 60% of the time. As a result, the reconnaissance party actually collected more specimens than the group who were mining our previously discovered concentrations, and that is not the way we had planned it. But then, things do not often go according to plan in Antarctica.

A BAD SUMMER, AND ITS CONSEQUENCES

During the period 1976–1995 we missed only one field season. This occurred in 1989, when the National Science Foundation (NSF) had an antarctic summer that can be described conservatively as very difficult, and more realistically as an organization's fever dream, during which a series of problems interacted in such a way as to magnify the total nasty effect. If the Fates conspire, they surely conspired that year against the NSF.

- This was a year of high sunspot activity, so that radio communications were often blacked out.
- The season saw more bad weather than normal, so that more flights had to be canceled.
- There was a higher incidence of aircraft malfunctions: the mechanics could not keep up with the repair schedule and often there were planes sitting on the ground, waiting to be repaired.

Dealing with radio blackouts posed particular problems. A plane coming down from Christchurch, New Zealand with supplies and passengers has to make radio contact with McMurdo before passing its point of safe return. If sunspot activity has shredded the ionosphere, this becomes very difficult. As a result, even if McMurdo is enjoying excellent weather, the pilot may have no way of knowing this and must turn back. There were times when they launched two LC-130s northward from McMurdo in order to loiter at stations spaced in such a way that they could relay line-of-sight messages between McMurdo and an incoming plane. Similarly, planes are not going to leave McMurdo for South Pole Station without checking first on the weather there, and there were long radio blackouts with South Pole.

As the season advanced, the situation worsened. There were air cargoes and passengers stacked up at McMurdo waiting to get to the south pole, and unable to leave because weather conditions at the pole were unknown. There were field people stacked up at McMurdo waiting to be put in at remote field camps, who could not leave because of bad weather at McMurdo or too many aircraft breakdowns. There were scientists stacked up in Christchurch, unable to reach the ice, who would not have had any place to sleep even if they could have gotten to McMurdo. And there were two field parties who had not yet left the US – ANSMET and Gunter Faure's group from Ohio

State University. Faced with an impossible situation, none of which was of their own doing, NSF personnel had to make some decisions. The most logical immediate action was to cancel the field seasons of the two parties that had not yet left the US, and that was why we missed our field season that year. I received the telephone call at 7 a.m. on the day I would have left for California on the first leg of my trip south.

The news was not without its hardships – Mario Burger, a Swiss graduate student, was already in Los Angeles. In order to stop him from continuing on to New Zealand, I first had to call his thesis advisor in Switzerland, to find out where in Los Angeles he was staying. Bob Walker, the Director of the McDonnell Center for Space Sciences at Washington University, St. Louis, who was looking forward eagerly to spending the field season in Antarctica with his wife, Ghislaine, emitted a plaintive cry of disbelief at the news. They had rented their house to an arriving scientist who would be visiting the McDonnell Center while they were gone. He had to find alternate lodgings for his visitor at very short notice. And John Schutt, our crevasse expert, was already in Christchurch. When the news reached him, he had been summoned three times already to board flights southward that had then been canceled. He was sitting in an LC-130 at the airport when his fourth flight was canceled. As he was disembarking the plane with the other disgruntled passengers he was told to see the NSF representative at Christchurch, who gave him the news that our project had been canceled for that year.

When a group of people spend from six months to a year preparing for and anticipating a once-in-a-lifetime experience, it comes as a shock to be informed on short notice that it has been postponed for a year. These people had been preparing themselves and their families for the separation; in many cases, they had mapped out their research plans with a three-month absence in mind, had arranged for others to take over their teaching schedules, had paid bills in advance, and who knows what else? They had been poised to go and needed time in which to reorient their thinking.

The problem was that when the news arrived we had no idea how serious the logistics situation was; therefore we were not equipped to understand what seemed to us to be a rather arbitrary decision. In our frustration we wondered if this was a first indication that ANSMET was being singled out for progressive de-emphasis. This fear also seeped into that part of the worldwide scientific community that had benefited by receiving antarctic meteorite samples for their research. They wondered if the supply was in danger of drying up, and resorted to the usual response mechanism used widely these days in trying to influence government agencies: they wrote letters explaining their research and the importance of a continuing supply of antarctic meteorites to that research.

The letters came from Austria, Canada, the former Czechoslovakia, Denmark, France, Germany, Great Britain, India, Italy, Peru, Switzerland, and the United States. By that time Mort Turner had retired as program manager for the Earth Sciences, and I think that Herman Zimmerman, who had replaced him, had to read all these letters and respond with a gracious form letter from Peter Wilkniss, who was then director of the Division of Polar Programs. In an exercise of this sort, there is a fine line between tolerance and resentment. There seems to be evidence that all this correspondence left a lasting and, importantly, a generally favorable impression on the Division of Polar Programs personnel concerning the value of the ANSMET project. The blizzard of letters is mentioned now and again at Meteorite Working Group meetings. This is a measure of the impression they made, and we can only hope that their corporate memory is not a bitter one.

THE RESULTS

During 24 field seasons from 1976 through 2000, the ANSMET project recovered more than 11 000 meteorite specimens. This number includes five putative martian meteorites and eight undoubted lunar meteorites, as well as many asteroidal meteorites that are unusual enough to excite special attention in the meteoritics community.

GROWING OLD IS NOT FOR SISSIES

There are things we notice about ourselves as we grow older, and most of them are not fun. It happens to some sooner than to others. I will not go into the long, sad list of ways in which we notice a progressive deterioration in our ability to function as before, and will name only one: blood circulation in the extremities. During the last couple of years during which I was Principal Investigator on the ANSMET project, I began to notice that my hands and feet got cold more easily than they used to. Coupled with this was a new tendency for the skin over my cheekbones and my nostrils to lose circulation. This is called frostnip – a condition that precedes frostbite. Frostbite is a very serious cold injury in which an affected part of the body freezes solid. Frozen cells die. You do not feel the onset of frostnip. Because of this, my colleagues in the field had to be sure to glance at me from time to time in order to warn me when the white, waxy-looking spots that herald frostnip would appear on my face. I would then have to stop whatever I was doing and vigorously scrub those places to warm them up and restore circulation.

This was a clear signal that my days working on the ice plateau of East Antarctica were nearing an end, and I should be thinking about finding a successor. I had a list of characteristics that I considered are desirable attributes in the ideal ANSMET principal investigator. These were characteristics either that I recognized in myself and had found to be valuable, or, more frequently than I liked, did not see in myself but wished they had been present. I will not go into details. Suffice to say that my first choice for a new leader of the ANSMET project was Ralph Harvey, who had done his PhD program at The University of Pittsburgh and with whom I had been in the field a number of times aleady. Luckily, Ralph liked the idea, so we submitted a proposal for continuation of the work with Ralph and me as co-principal investigators. Ralph is now, of course, running the operation on his own. He has been very successful and I believe he was the best choice I could have made. My last season in the field with the ANSMET team was the austral summer of 1993–94.

5 Alone (or in small groups)

Under certain circumstances, being completely alone can get to one. Being alone in Antarctica can get to one pretty fast. Admiral Byrd wrote a book entitled, *Alone*. Even being part of a small group, one experiences a sort of group aloneness, leading to a sense of awe, perhaps, at the total isolation of this small nucleus of humanity whose individuals are completely dependent upon one another, not only for intellectual stimulation but even for nourishing the basic need of the mind to feel that it is still a part of the fundamental structure of human society. We deal with the sense of isolation through conversation, clowning around (Figure 5.1), and seizing upon the arrival of holidays to hold parties. But it is definitely an abnormal existence, during which strange and memorable things sometimes happen. Following are some events that seem to fall into that category.

EVENT I: WORKING ALONE

In the field we try never to be alone, usually working in pairs or as a complete party. The practical rationale for this rule is that one person, alone, can get into a lot more trouble with crevasses, or stranding due to mechanical breakdowns, than would two people. Traversing between campsites is potentially the most hazardous operation if we are taking a path for the first time. There may be crevasses. At such times the field party travels as a unit. To deal with the question of crevasses we have a "crevasse expert." The term defines the person who goes first, and he had better know his crevasses, to avoid falling into one, and rescue techniques in case someone else goes into one. All our other field personnel are unpaid volunteers; the crevasse expert is the only one who receives a salary, in recognition of his special

FIGURE 5.1 Carl Thompson steps out of his tent to make a wry comment about the "one size fits all" nature of the long underwear he was issued. (Photo by C. King-Frazier.)

abilities. But he must return safely, because he doesn't get paid until the field season is over.

We have had an accomplished alpinist named John Schutt as our crevasse expert during most of the seasons I have spent in Antarctica. John is of medium stature, but in Antarctica he stands "10 feet" tall. He has tremendous upper body strength. His recreation is mountaineering. When I was seeking a crevasse expert to hire, I heard the following story about him.

John and a friend were traversing a knife ridge during a dual climb. They were roped together for safety. John's friend, who is about twice John's size, was leading. Without warning, he fell off the ridge (these things happen). As he slipped, he was suddenly looking straight down about 300 m. John immediately realized he would be pulled off his perch, too, so he launched himself into space on the other side of the ridge. It was the only chance he had to brake the other's fall, and it worked, saving both their lives. We didn't plan to climb any mountains, but I knew I wanted John Schutt out there with us.

Only rarely do we ignore the buddy-system rule. This usually happens around our campsite or on terrain that we have traversed many times and that we "know" to be safe. During our 1983–84 season we visited, for the second time, the meteorite stranding surface at

Elephant Moraine (see Figure 4.5). There are a couple of areas around Elephant Moraine where crevasses can be found, and we all knew to stay away from those spots. The rest of the surface was exposed ice, where any cracks or crevasses are quite visible. If anyone goes down a crevasse on bare ice, he deserves it.

We had finished a day of mapping at Elephant Moraine and were back at camp, when it occurred to me that we could have picked up our marker flags as we went along, and this would have saved time on the following day. I told John that I would go back and collect them all, and rather than impose on any other tired person, I would go alone. It took about an hour and a half to collect most of them, and this was longer than I had expected. The aloneness was beginning to get to me. Antarctica is a desolate place by anyone's standard except, possibly, that of a Martian. I had been out of sight of camp for a long time, the wind was kicking up, it was past my dinner time, and unaccountably I started to feel nervous. I got a vague feeling that there was someone out there with me, and looked over my shoulder a couple of times. Suddenly, shockingly, there *was* someone – just in front of me. John thought I had been gone too long and had come to look for me. I greeted him with a great sense of relief.

EVENT 2: WHEN IT GETS QUIET

During one field season near the Allan Hills in Southern Victoria Land, our field party of six had been working at a meteorite stranding surface called the Allan Hills Far Western Icefield. This location is about 90 km west of Allan Hills, which is the closest rocky exposure. We were making our way toward Allan Hills and constructing a survey line as we went, in order to be able to tie in the Far Western Icefield to fixed points on rock at the Allan Hills. Our survey turning points were taken about 1500 m apart – about as far as John Schutt could manage with the theodolite and infra-red ranging device. He would set up the instruments on a new station, back-sight at the previous station, where one of our party had been left with a set of infra-red reflectors, and swing around to locate a forward point where an advance member of the party would have set a flag. Then we would all move to the forward point and repeat the operation, working our way slowly toward Allan Hills, inadvertently building a slight cumulative error into the location and relative elevation of each station. This error would be distributed equally among the stations when we had finally established our real position from known points on the solid rock of Allan Hills.

Bob Fudali, of the Smithsonian Institution's US National Museum, was taking gravity readings at each of our stations. In the absence of radio echo-sounding equipment, gravity data can lead to next-best estimates of ice thickness. This, combined with elevation readings, can give an idea of the under-ice bedrock topography. Bob is extremely conscientious about the quality of his data, and his gravimeter had given a couple of readings that he mistrusted. He decided he needed to redo these measurements and would have to go a couple of stations back to repeat them. He informed John Schutt and started off on his snowmobile, along our backtrack. I checked with John to see what was happening, and decided to go back with Bob so he would not be alone. I swung into his tracks and chugged along for about a mile before noticing that he had not looked back and apparently didn't know I was following. He thought he was alone. We passed the first station and headed for the second, barely visible on the horizon. I decided to carry this as far as it would go. As we approached the next station he appeared to be planning to swing around. To avoid being seen, I swung wide and speeded up, to remain always behind him as he turned his snowmobile to head in the direction for his return trip. He had not seen me.

Snowmobiles saturate their immediate environment with noise. I parked just behind him without his seeing me and switched off my engine before he had done so. He switched off his engine and a sudden, intense quiet enveloped us. It was one of those rare days on the ice plateau of East Antarctica: completely windless with bright sunshine, and by contrast with the usual conditions it gave the impression of being warm. It was a magic moment. The only thing I can compare it to is the sensation a city dweller gets on those rare instants, when by some statistical anomaly, there is suddenly no traffic in the street for a four block interval. The roar of traffic blocks out more subtle sounds, and the city dweller blocks out the roar. When the roar unexpectedly and completely disappears, one is suddenly aware of all the minor sounds of the city that he may never have heard before. His own footfalls. A distant laugh. A grocery store clerk two blocks away tossing garbage cans, and the time lag for the sound to reach the observer. A magic moment!

Bob searched the horizon for our field party, but the dots had disappeared. The blue, cloudless sky formed a perfect hemisphere and the white snow, featureless except for shadows cast by its surface irregularities, stretched away to the sky in all directions. He did not know I was there. His sense of being completely alone must have been flooding in, along with the almost suffocating sound of silence. Very spooky. Then I commented in a loud voice,

It sure is quiet out here, isn't it!

We have never discussed this incident. He had nerves of steel, pretending he was not startled by a disembodied voice behind him, but his involuntary flinch gave him away.

EVENT 3: YOU CAN'T PUSH ON THE END OF A STRING AND EXPECT MUCH TO HAPPEN AT THE OTHER END
During our second field season, Billy Glass brought a level of intensity to the field work that I had not yet learned myself. In Bill's thesis research at Columbia University he had found certain horizons in deep-sea sediment cores that were marked by unusual numbers of microlife extinctions and the simultaneous appearance of new species. This work, carried out during months of staring through a microscope, had honed his observational skills to a fine edge. He brought this sharp eye to the field and turned it to spotting meteorites at a distance. He was very successful at this, and including him in the field party had been an excellent decision.

During his thesis work, Bill became the discoverer of micro-tektites – tiny glass spheroids buried in the sediments covering perhaps 10% of the ocean floor. In their shapes and compositions they are related to larger glass objects called tektites that are found on land, and are believed to be melt droplets from titanic impacts between asteroids and the earth. I had been interested in tektites for years, and we often had long discussions in the tent on these and other subjects.

One evening, Bill mentioned something that ordinarily he might have kept to himself. He said he was worrying considerably about a letter he had written to his wife, Judy, just before leaving McMurdo for our field stay. In it, he had complained of her recent decision to start smoking, even though she knew he disapproved. He was worried that he had been too judgmental, and worried, also, about how she might react. And now here he was, out in the middle of a very cold Nowhere pushing on the end of a string, completely out of touch; unable to get any conciliatory message home; dreading to find an answer to his letter waiting at McMurdo and dreading even more that there would be no answer when he got back. Actually, there was no letter when we returned. It came in a few days, though, in the form of a standard file card in an envelope. This remarkable woman had needed only four words to convey volumes.

> STOPPED SMOKING.
>
> STARTED DRINKING.

EVENT 4: INCOMPATIBLE FIELD PARTNERS

It is a known fact that perfectly nice people, who would be able to coexist amicably under almost all circumstances in ordinary life, can crack and come apart at the seams when exposed to conditions in a remote field camp in Antarctica. Requirements for survival are severe. People are forced into close and continuous contact with their colleagues. Field partners are few in number, and after a while cannot come up with fresh topics of conversation. All these factors create stresses that are not present in civilized society, and field personnel must learn to adapt, or crack.

This story was told to me by a source I consider reliable, but I was not there, and cannot vouch for the details. I have never known the names of the principal players.

Some years ago, the National Science Foundation (NSF) agreed to support field studies at a remote mountain range in Antarctica that was very interesting to a number of geologists and paleontologists. A large operation was mounted in which LC-130 aircraft ferried in three helicopters and all the necessary structures, supplies and support personnel for a remote base camp. The camp was established, the scientists arrived, and were deployed here and there by helicopter, up to distances from the base camp of 185 km – the maximum range allowed for helicopter flights. Some groups commuted to and from field sites every day by helicopter and other groups were put in to remote field camps, where they lived in tents while carrying out major geological studies.

A senior scientist took two graduate students to the field as assistants. They established a remote field camp high in the mountains and about 180 km distant from the base camp. They were about as far away from any other human beings as it would have been possible to be. Toward the end of a projected four-week stay the scientist was called back to the base camp, and he had to leave his students behind to finish the mapping.

Stresses had been building between these two unfortunates for three weeks, and when the helicopter with their boss aboard had dwindled to a dot in the distance they considered the isolation of their position and contemplated with distaste the necessity of coexistence for an additional week. They looked at each other, and cracked.

The basic problem and the cause of their animosity was that one was an absolute neatnik and the other lived in a perpetual state of complete disorder. They had been sharing a tent. They agreed that each would occupy his own tent for the remaining week. They hastily threw up the tent their professor had used and divided their belongings between the two tents. On the second night, each was in his own tent and waves of hate were radiating back and forth between them.

They went to sleep, and during the night a violent storm blew up. The untidy Student's tent blew away because he hadn't bothered to stake it down. He sprang up from his sleeping bag, practically in the nude, and immediately began to freeze. He stumbled over to the other tent and tried to crawl in, but it was tied on the inside. The Scott tents used in the US Antarctic program have an entry tunnel of canvas. When one leaves the tent the tunnel is tied shut from the outside; when one is in the tent, the tunnel is tied shut on the inside. He couldn't get in. In a panic, he began to shake the front of the tent, screaming, "Let me in! Let me in!"

After a prolonged interval, a voice inside the tent said,

Who is it?

EVENT 5: A DAREDEVIL HELICOPTER PILOT

In 1976, the first year of the ANSMET project, and for four or five years after that, all the helicopter pilots at McMurdo were Vietnam War veterans. These remarkable men had survived extreme dangers on a routine basis. They had been called upon to put troops in and extract them from some very risky situations and had survived, apparently by learning every nuance of performance that these aircraft were capable of and becoming an organic part of their helicopter during flight. They formed a pretty closed society at McMurdo. At remote base stations they also kept to themselves. They always took Sundays off, and every Sunday one could find them over at the helicopter pad, leaning against their machines and idly conversing or joking around (Figure 5.2).

These Vietnam-era veterans took a relaxed attitude toward the existing set of rules and regulations concerning their activities. They knew their aircraft. If their helicopter was capable of a maneuver, and

FIGURE 5.2 Sunday afternoon activities for the helicopter pilots.

they knew it was capable, and it seemed worthwhile to do it, they
did it.

During these years, when being ferried back and forth between
McMurdo and one or another field site on the other side of McMurdo
Sound, it was a common occurrence for the pilot to set his helicopter
down near the ship channel. The ship channel is opened by a US
Coastguard icebreaker every year so that a small cargo ship and a small
oil tanker can get from the ice edge to McMurdo with supplies for the
following 12 month period. The channel immediately becomes an
interesting avenue of exploration for those large, air breathing marine
predators, the orcas – killer whales. These guys travel in pods of two or
three, breaking the surface together and diving again in unison. Often
they float up to a nearly vertical position, and are buoyant enough
that their entire head and upper body, including relatively tiny front
flippers, are out of the water. They can remain this way for some
time.

Why do they do this? They are scanning the ice surface for penguins or seals. If the ice is thin enough they rise up under their prey,

breaking the ice and throwing dinner into the water where it is speedily consumed. Observers report that they often cooperate in this ice breaking activity, as two or three rise simultaneously under their unsuspecting quarry.

At the edge of the ship channel one can get closer to killer whales than at any other location. If the ice is still thick enough to support the weight of a helicopter it is thick enough so that people standing on the ice are in no danger of becoming a meal for a killer whale who thinks he is getting a particularly tasty penguin. Up close and personal, at the edge of the channel, one can look an orca in the eye as he floats up in his vertical mode and looks back in an appraising sort of way, with his mouth open in a terrible grin, displaying rows of very businesslike-looking teeth.

The helicopter pilots saw no particular danger in setting down on the ice by the ship channel, and foresaw an opportunity for great photography, so they often did it. Their cargo of scientists always was equally appreciative of this opportunity.

My most memorable experience along these lines was an occasion when we were returning from Allan Hills and the senior pilot was one Lt. M. B. (M. B. stands for Mick Brown. I use his initials only to avoid disclosing his real identity. Oops!). Mick spotted some pods of orcas along the ship channel. He set us down some 30–40 m from the edge, and we were able to walk over and get some photos. The channel was full of large, floating chunks of ice. Some of these had become frozen to the ice edge and extended out into the channel, where several pods of killer whales were repeatedly breaching and diving, like synchronized swimmers in an aquacade. Mick crawled out onto one of these frozen chunks, took the gloves off his left hand, and waited. Soon a pod of orcas came up next to his position and as they dived he reached out and *grabbed one by the tail*. This animal was 4–5 m in length. Mick couldn't hold on and the tail slipped from his grasp as its owner dived. Mick scrambled back onto the relative safety of the main body of ice and walked back to his helicopter, immensely pleased with the experience and wearing a broad grin. He reported that the orca gave no sign that it had felt anything, and that the tail felt like wet rubber. He said, simply,

I have always wanted to do that.

In the isolation and social deprivation of these field situations, people will tell you more about themselves than they would under more normal circumstances. Often the results are surprising, revealing unsuspected depths of character or experience. If I had to pick that one person who surprised me the most, it would be Paul Pellas.

AN APPRECIATION OF PAUL PELLAS

Paul Pellas, of the *Musée National d'Histoire Naturelle,* in Paris, was a member of the 1983–84 ANSMET field team. In any field project such as this, in which many people from many countries have taken part, some will always stand out in the memory more than others. Usually there are anecdotes that accompany these memories. Paul Pellas's life, however, seemed to be one continuous anecdote.

During World War II his country was occupied by the German army, and in the early 1940s he joined the Communist Party. I was surprised to learn this, because Paul did not seem to adhere to a particularly strong political ideology. As he explained it, most of the Resistance groups in France concentrated on sabotage but the Communists exerted most of their efforts in helping wanted persons, mainly Jews, to escape to a safe country. He thought this was worth doing, and was involved in a number of such escapes, leading groups across borders at night, sometimes on skis.

For a while after the war he earned his living as an actor in supporting roles, once playing in a movie whose star was Brigitte Bardot. We asked him repeatedly, "Whether he ever ...?" But he always maintained a gentlemanly silence.

He became a petrologist and worked at the National Museum of Natural History in Paris on the mineralogy of deep-seated rocks and meteorites. Because of his Communist past, he could not get a visa to visit the United States until 1970. In 1969, we had collected samples from the moon and he had been sent some for his research. Coming up in January of 1970 was the first-ever Lunar Science Conference in Houston, Texas, at which the first papers describing samples from the moon would be presented. This was to be a historic occasion and it was being argued that Paul had a right to be there. The State Department agreed to issue him a visa, as a participating scientist.

Paul arrived at Houston on an international flight that was about 50% occupied by fellow scientists planning to attend the same meeting, and faced US Customs for the first time. The fellow in the line ahead of Paul happened to have bought a Camembert cheese in France, which he was showing to the customs officer. This worthy minion must have thought it was about the right size and weight to contain a kilo of heroin, because he produced a large needle and began to probe the cheese. Paul had no idea who the fellow in front of him was, but he was outraged at such treatment of a Camembert, and immediately protested to the customs officer, waving his arms and shouting, "Stop! Stop! You will kill the cheese!" The customs officer must have thought he was about to catch a drug smuggler, because he then attacked the cheese with renewed enthusiasm and Paul became ever more distracted.

Suddenly two men wearing conservatively tailored suits and unsmiling faces appeared at Paul's side. They asked him to step out of line and accompany them. He protested to them about the cheese but they were adamant, and he was marched into a nearby room. As the door closed, he was still protesting the rape of the cheese in a loud voice. As soon as his friends had cleared Customs, they phoned acquaintances at the Johnson Space Center. NASA sent a public relations person out to the airport to negotiate Paul's release, which by then must have come as quite a relief to a number of federal agents. The fellow with the cheese, who had *not* protested what the agent was doing, had learned a lesson of somewhat limited applicability but which was, nevertheless, very precise. Never stand in line ahead of Paul Pellas, if you are trying to bring a cheese into the country.

Paul did love the good things in life. When I was assembling the provisions list at McMurdo for the 1983 season, I asked everyone for their preferences in alcoholic beverages for use in the field. When it appeared that most were favoring strong drink, Paul made an impassioned plea for wine. He had tears in his eyes. "Remember his acting experience," I told myself, but finally agreed to consider the possibility. I pointed out that wine would freeze, and asked if he really wanted to subject a delicate blend to such a brutal process. "We will not use French wine," decreed Paul, ever the patriot. "Also, we can run some tests ahead of time to see if it is practical. We will try only white wine." So I bought a couple of bottles of white wine and we drained part of each bottle, to keep unfrozen in polyethylene bottles. The remaining partial bottles of wine were stored for several days in a freezer and then thawed out.

There was a slight residue of white powder inside the bottles when the wine was thawed, but in comparing the frozen wine with the unfrozen, our only observation was that the frozen wine emerged from the process slightly drier than before. So we went to the field with wine, but not before draining off, and drinking, about 10% of each bottle in order to allow room for expansion during freezing.

Our goal in the 1983 season was Elephant Moraine, and we had arranged to be put in about 40 km away by an LC-130. After landing, we were intensely busy unloading cargo and setting up the radio antenna because we would not be allowed to stay, without having demonstrated our ability to communicate with McMurdo by actually exchanging messages. Off to one side was Paul Pellas, calmly uncorking each bottle in turn, in order to let the contents adjust to the atmospheric pressure at our 2000 m elevation. We drank them with appreciation on Christmas Eve and New Year's Eve, 1983, while consuming hors d'oeuvres of pâté de foie gras that Paul had carried all the way from Paris.

Paul was fastidious to a fault. In Christchurch, New Zealand, we were issued mainly used, but cleaned, clothing to wear in Antarctica. The sleeping bags, however, were issued at McMurdo. They are aired out at the end of a season and every couple of years are shipped to Christchurch for dry cleaning. Paul was tenting with Ludolf Schultz, and after the field season Ludolf told me that the first night in the tent, Paul had crawled into his sleeping bag and then immediately crawled out again, with the comment, "*It has not my odeur.*" For the entire field season, therefore, he slept only halfway into his sleeping bag, with his parka over the upper part of his body for warmth. "It has not my odeur" became a common complaint during succeeding field seasons, usually applied to something a tent-mate had fixed for dinner.

Paul died during the last week of May, 1997. He is missed.

Part II ANSMET pays off: field results and their consequences

I am lucky to have induced my colleague, Bob Fudali, to review early drafts of each chapter of this book. One of Bob's major talents is a fine critical faculty. Another is a mindset that allows him to speak without fear or favor. I recently received the following note from him (tucked into a Christmas card).

> On several occasions I've gently tried to tell you that you are seriously overestimating the importance of the vast majority of the meteorites found in Antarctica. That is not to say we shouldn't continue to collect them – only that we should not become overly enamored of our own importance (as reflected by the "singular importance" of the antarctic meteorites). I will, of course, continue to bring this to your attention.

A characteristic typical of most of us is to be overly impressed with the value of our own research, and I do not claim immunity from this failing. If I give an impression of excessive pride in the following chapters, even after Bob's stern admonition, then I apologize, but let me also point out to the reader that he/she has been warned...

The total of all meteorite specimens recovered from Antarctica by US, Japanese and European teams now numbers somewhere in the neighborhood of 30 000. Among these are a few lunar samples that fell as meteorites after being blasted off the moon as a result of collisions with asteroids. Included also are a few samples that we believe are from Mars, derived from the martian surface by the same mechanism as the one that produced the lunar meteorites. The great majority of the collection, however, consists of fragments of asteroids – bodies that inhabit orbits between those of Mars and Jupiter. Most asteroids occur in a belt more than twice as wide as the distance between the earth and the sun. The asteroid fragments that we receive as meteorites, both on the antarctic ice sheet and on the rest of the earth's surface, display a wide range of compositions and textures, suggesting that they come from a large variety of parent bodies. These range from bodies that may have been completely melted above 1200 °C and then crystallized, through bodies that may have been only partly melted, to possible low-temperature accumulations or even comet nuclei that have never been above 200 °C. We have not yet been able to send spacecraft to sample asteroids directly, therefore part of our fascination with meteorites resides in attempting to reconstruct the compositions and evolutionary histories of their parent bodies. We thus hope to enlarge our knowledge of the varieties of orbital objects that populate the asteroid belt. Beyond this, however, is the possibility that we can recover fragments of asteroids that no longer exist as large bodies, having been diminished by minor collisions and/or crushed and scattered during major collisions. Fragments produced in this way have a good chance of being thrown into resonance orbits relative to Jupiter and Saturn. In a short time (as geologists measure time), these fragments are set into paths out of the solar system or into paths that will result in collisions with Mars, Earth, Venus, Mercury or the sun. It is probable, however, that some fragments of almost all asteroids that ever existed in the asteroid belt still remain there, but that they are small and will never be visited by spacecraft. The laws of physics being what they are, a small fraction of these will rain onto the earth, where enthusiastic gleaners, particularly on the antarctic ice sheet, will collect them and be able to add more bits to a paleohistory of asteroidal bodies.

I have constructed Chapter 6 to deal with martian meteorites, and Chapter 7 to discuss only lunar meteorites. In Chapter 8, I attempt to summarize all the others.

In Chapters 6–8 I try to give a sense of history to the controversies surrounding the martian and lunar meteorites, and to the classification of asteroidal meteorites, while also including a summary of what these meteorite types may be telling us. In general, I have attempted to convey a sense of the great variety among known meteorite types and to indicate how the enormous, and still growing, collection of antarctic meteorites has changed the numbers of samples of each type available for study. Readers might wish to refer to selected references that I include at the ends of the chapters.

I am trying to write for the intellectually curious general reader. Hopefully, by the end of Chapter 8, one will have seen the name of each meteorite group at least once and they will seem less like words and symbols in a foreign language. For those unfamiliar with the terminology, learning will have begun. Welcome, you intellectually curious general readers!

6   Mars on the ice

INTRODUCTION

A mountain, a mesa, a cliff and, in fact, any major outcropping of rock normally has an accumulation of weathering products piled up around its base. Such an accumulation of rocks is commonly called *talus*. When these rocks are carried away from their site of origin by river or glacial transport, they become *float deposits*. As geologists, we prefer to collect our samples directly off the outcrop because we can precisely locate where on the outcrop they originated. If the outcrop is a steep cliff that we cannot climb, we may settle for a collection of rocks from the talus. Rocks collected from float deposits, however, may have been sorted during transport and may derive from many different sites along the path of the transporting agent. These contain the least information value and are generally ignored in favor of the other two sources. But suppose the original outcrop had been destroyed and its talus carried away, or for some reason was inaccessible? By now, geologists would have figured out all sorts of clever ways to extract information from the float about the inaccessible outcrop from which it came. This, of course, is the situation faced by meteoriticists: a meteorite is a sample of the detritus from an outcrop that either is inaccessible to us, or that no longer exists. A further complication is that in the aggregate, meteorites must represent many different outcrops, from many different bodies in the solar system, and we are called upon to sort these out, devising ingenious ways to study them in a meaningful manner. So, historically, we have searched the world in order to accumulate as many separate meteorite falls as possible, hoping thereby to come to know the full range of unearthly environments within which these rocks formed. Only recently, however, have we begun to reach for meteorites in Antarctica where, as we all know by now, astonishingly high *concentrations* of them can be found on certain patches of ice.

One of the major results from the ANSMET field project has been the recovery of meteorites believed to be samples of the planet Mars. We know that ice occurs on Mars; now, we know that parts of Mars occur on the ice in Antarctica. Someone who knows very little about meteorites, hearing this for the first time, might have a couple of questions, such as: "How do you know these rocks came from Mars?," and "If they are from Mars, do they carry evidence of life on that planet?" Someone who knows a lot about meteorites, hearing these questions, would recognize the validity of the first one and, until very recently, chuckle good-naturedly over the second. Recent research, however, suggests that the second question is also a valid one, and our naive friend, quoted above, has gone directly to the heart of the matter.

NON-ANTARCTIC MARTIAN METEORITES

The first meteorites to be suspected of having a martian origin were not antarctic meteorites at all. Well before the ANSMET program, the clan of suspected martian meteorites had three families called shergottites, nakhlites and chassignites. These are names that would not have come out of Antarctica. Shergottites are named after the first identified member of this group, a 5 kg meteorite that fell near Shergotty, India at 9 a.m., August 25, 1865. Subsequently, an 18 kg stone fell near Zagami in Katsina Province, Nigeria on October 3, 1962. Even though the major minerals making up Shergotty had crystal sizes twice that of Zagami, the minerals of Zagami and their compositions were so similar to Shergotty that it was classified as a shergottite.

A shower of at least 40 stones weighing about 10 kg, total, fell near the village of El Nakhla El Baharia, about 39 km east of Alexandria, Egypt at 9 a.m. on June 28, 1911. It was reported in the newspapers that one of the stones killed a dog! Suppose this story were true...Imagine an asteroid orbiting the sun in the asteroid belt for

4.5 billion[1] years. Meanwhile on Mars, fractional melting beneath the surface produced a body of magma, some of which was emplaced near the surface as a shallow intrusion and solidified about 1.3 billion years ago. The destiny of the asteroid was to collide with Mars 11.5 million years ago (give or take a couple of million years), impacting the planet's surface in just exactly the right way as to blast pieces of that solidified magma off the surface of Mars. One of those pieces left with that dog's name on it. This piece would escape the gravitational field of its planet and assume an orbit about the sun that eventually would take it near to the earth. On the earth, canine forms appeared in the geological record around six million years ago, and began to evolve in the direction of man's best friend, the noble dog. Meanwhile, perturbations were occurring in the orbit of the rock from the surface of Mars that would converge with the evolutionary history of *Canis*, coming to a point in space and time on the surface of the earth at 9 a.m. of June 28, 1911, when the Martian rock entered the earth's atmosphere, breaking into smaller pieces in just the right way so that one of its fragments conked a particular dog, who had taken just a few steps in the wrong direction. Talk about bad luck! That dog had it before he was ever born and even before his genus had evolved! The meteorite was called Nakhla. History does not record the name of the dog.

Left to the philosophers is whether we should look at this as a story about the interconnectedness of all things, or a story about "... the slings and arrows of outrageous fortune..."

Two other non-antarctic meteorites of the nakhlite class are known. Lafayette is a very poorly documented stone of about 800 g that must be a very recent fall, judging by its perfectly preserved complete fusion crust. It was found in 1931 in a geological collection at Purdue University in West Lafayette, Indiana and is believed to have been picked up in Tippecanoe County. Governador Valadares is a fresh-looking stone, and therefore probably a recent fall, weighing 158 g, found in 1958 in Minas Gerais, Brazil.

---

[1] Billion years = thousand million years.

Chassignites are named after a stone seen to fall at Chassigny, France at 8 a.m., October 3, 1815. One stone, weighing about 4 kg, was found.

Together, the shergottites, nakhlites and the lone chassignite came to be known as the SNC (pronounced snick) meteorites. What did these three families of meteorites have in common, that they should be grouped together? Three things, apparently: (1) they are all igneous rocks; (2) they all had similar oxygen isotope ratios; and (3) they all had crystallization ages of 1.3 billion years, or less. These young ages of formation distinguished them from all other igneous meteorites, which formed around 4.5 billion years ago, or within about 100 million years after the formation of the solar system. Where in the solar system could the SNCs have crystallized so recently from a melt?

A PROCESS OF ELIMINATION

The earth has generated igneous rocks throughout its history, and igneous rocks are still being formed today. The oxidation state under which the SNC meteorites formed is about the same as that of terrestrial basalts, and the SNCs themselves do look remarkably similar to terrestrial basalts. Could the SNC meteorites be terrestrial fragments produced by gigantic impacts on the earth, millions of years ago? A rock thrown off the earth in this manner would have to have been inserted into a solar orbit and, after a million years, or so, during which it acquired a history of exposure to cosmic rays, fell back onto the earth, allowing us to reacquire the rock as a meteorite. While it cannot be proved that such a history is impossible, the velocity of escape from the earth is 11.2 km/s, so an object would have to exceed this velocity or it would immediately fall back. Escape velocity from Mars is about 5 km/s – less than half the earth's, so at least it would be much easier to boost a rock off the surface of Mars than of Earth.

The earth–moon system cannot have been the source. Robert Clayton, at the University of Chicago, has measured oxygen isotope ratios in hundreds of meteorites, as well as lunar and terrestrial rocks.

He finds that these rocks can be separated into groups that presumably have different parent bodies, or at least come from different regions of the solar system, depending upon their $^{17}O/^{16}O$ and $^{18}O/^{16}O$ ratios. SNC meteorites plot in a group separate from that of the earth–moon system, and separate also from other groups of meteorites. This illustrates the unique nature of the SNCs and also rules out the earth–moon system as their source.

Asteroids are too small to have been their source. We believe that most of the meteorites that fall on the earth, even the meteorites of igneous origin, formed in bodies that were much smaller than the moon. We can see how igneous rocks could have formed in magma chambers early in the history of such a small parent body before the heat generated by gravitational accretion, core formation, short-lived radioactive decay and compression had dissipated. But it would require a substantially larger parent body, even larger than the largest asteroid, Ceres, which has a diameter of 1020 km, to retain sufficient heat, sustained over time by long-lived radioactive decay, to continue to generate magmas for at least 3.5 billion years after it had formed. The SNC meteorites apparently came from a larger parent, and asteroids are too small. Confirming the magma chamber source is the fact that igneous minerals in some of the SNCs showed preferred orientations. There are only a couple of ways such a fabric can be induced: it can be produced in a magma chamber by slow crystal settling in a cooling magma, or by parallel orientation of crystals in a flowing magma, shortly after extrusion from a magma chamber.

Io and Mercury are less likely sources than Mars. Of the planets and larger moons in the solar system, only the earth and Io are known to have volcanic activity currently, although Venus may. The earth has been dropped from consideration because its rocks have the wrong oxygen isotopic ratios. Rocks blasted off Io would be more likely to be injected into orbits about Jupiter, rather than about the sun, and the dense atmosphere and high escape velocity from Venus make it a less likely source of ejected rocks than the smaller planets such as Mercury, which has no atmosphere, and Mars, with its very tenuous

atmosphere. The surface of Mercury appears to be too ancient to supply rocks only 1.3 billion years in age. An apparent deficit of impact
craters on the slopes of the large volcanoes on Mars, however, indicates that many of these volcanoes are rather recent. Thus, Mars
became the probable candidate for the source of the SNC meteorites.

Today, with the benefit of hindsight, one could believe that this
reasoning should have biased everyone toward the idea that SNC
meteorites come from Mars, but this was far from the case. There
were, however, hints. In 1975, Arch Reid and Ted Bunch, comparing
Lafayette and Nakhla, and by implication discarding any idea that
they were terrestrial, nevertheless concluded that,

> ... the position of the Nakhla parent body relative to other
> planetary bodies remains to be established. It is unlikely to be that
> of the more common chondrites or achondrites, and it apparently,
> in some respects, resembles the Earth.

So we had to look for a planet that, "in some respects, resembles
the Earth." In common with many others, their imaginations were
stimulated by the report that Nakhla had killed a dog. They close by
saying,

> Nakhla is the only meteorite for which there is evidence that its
> formation was contemporaneous with the presence of moderately
> evolved life forms on Earth. It is also the only meteorite known to
> be involved in the destruction of a highly-evolved terrestrial
> organism. The symbolism of this low probability correlation is
> obscure.

In 1979, Edward Stolper and Harry McSween published a paper
comparing the SNC meteorites to the other basaltic achondrites (i.e.,
igneous meteorites that resemble terrestrial basalts). They pointed
out that the SNCs were closely related to each other but different in
various ways from other basaltic achondrites, and probably derived
from a different parent body. They both believed this "different parent body" was Mars, but they very carefully avoided saying so. They

were, after all, very junior members of the community of planetary scientists, and it is a measure of the controversy they saw themselves becoming embroiled in, that they chose not to invite disaster. The actual proposition first appeared in print in a paper by John Wasson and George Wetherill in 1979, two scientists whose senior standing in the community provided some degree of insulation from gratuitous criticism. Even so, their wording was cautious:

> When combined with its relatively small atmosphere and gravity field, conditions should be more favorable for ejection without melting from Mars than from the earth.

and, later,

> Although it is obvious from the foregoing discussion that there are serious difficulties which must be surmounted before a Martian origin for these meteorites appears likely, the possibility that we have rocks from Mars in our meteorite collections is important enough to warrant giving some attention to this consideration.

Also in 1979, seven authors at the Johnson Space Center, led by Larry Nyquist and Donald Bogard, discussing the "shergottite parent planet," said,

> Early differentiation plus late magmatism imply a sizable parent planet. A portion of the planet's surface must be young. Mars fits these criteria.

In 1980, McSween and Stolper, after disposing of some of the other possible sources for the SNC meteorites, finally took a stand with the following comment,

> The only alternative source of the shergottites that we find appealing is Mars.

There it was, finally!

PROBLEMS WITH PHYSICISTS

Even though Mars appeared to be the most likely source of the SNC meteorites, the scientific community was told that collision with an asteroid could not possibly knock fragments off a body that large. Initial calculations suggested that the energy density (i.e., the amount of energy injected into a small volume) required to project rocks the size of these meteorites off the surface of Mars and into orbits about the sun would be sufficient to pulverize and/or vaporize the rock, so that it would not survive as a rock. While laboratory impact experiments tended to support this conclusion, they also have shown that if an impact is highly oblique, impactors traveling around 5 km/s can produce jets of at least partially unmelted target fragments, some of which may have the velocity of the impactor. Velocities of many Mars-crossing asteroids would be greater than escape velocity from Mars, so it seemed likely that an asteroid on course to a grazing encounter with that planet might actually confound the calculations of impact dynamics by imparting greater than escape velocities to a small fraction of the total volume of martian near-surface material affected by the collision, while at the same time preserving its original rocky nature.

With an ever-growing conviction on the part of planetary scientists that SNC meteorites actually do come from Mars, those physicists showing that this is impossible according to the calculations of impact dynamics found themselves in an uncomfortable position between a rock, represented by the laws of physics, and a hard place provided by mounting evidence that these meteorites indeed are pieces of Mars.

If the SNC meteorites came to us as impact products from Mars, however, there was another problem. The earth's moon is much closer to us than is Mars, and Earth is therefore a much larger target to receive lunar impact products than it is for martian fragments. Why, then, had we found no *lunar* rocks that had fallen as meteorites onto the earth's surface? So the planetary scientists also found themselves between a rock, represented by the physicists, and a hard place imposed by the missing lunar meteorites. This was a very troubling question until we began to find lunar meteorites also.

Shortly after we found our first piece of the moon that had fallen as a meteorite, Jay Melosh, at the University of Arizona, reexamined his impact model and discovered a zone around the outside of an impact crater, two or three projectile radii out from the center and near the surface of the impacted body, where interferences between shock waves could be expected to impart very high velocities to spall fragments, while at the same time shocking them very little, or not at all. In 1984, with the usual disclaimer, he wrote:

> A great deal remains to be done. It is quite clear, however, that *some* intact rock material is ejected from planetary parent bodies by large impacts. The process of stress-wave interference, a minor aspect of the main cratering event, is of inordinate interest for geologists because it marks the beginning of a meteorite's obscure journey from parent body to terrestrial museum.

To a geochemist, of course, a meteorite's journey is anything but obscure; particularly that part of it between being collected and being finally archived in a "terrestrial museum." In that short time interval, literally hundreds of scientists have the opportunity to wring from it, if not secrets of the universe, at least secrets of the solar system and, in that short part of its journey, remove it forever from the obscurity to which others might carelessly consign it.

ANSMET FINDS A MARTIAN METEORITE

During December 1979–January 1980 (we call this the 1979 field season) four of us had traveled by snowmobile up to a large ice patch just west of Reckling Peak, where the Phil Kyle party had found five meteorites the year before. Getting used to the idea that *concentrations* of meteorites can occur in Antarctica, we reasoned that where you can find five meteorites, there ought to be many more. We quickly collected 15, and this helped to establish the principle we have followed since then: if you can find a few meteorites quickly on a given patch of ice, then you have located a new stranding surface and it will pay to carry out systematic searching. In later seasons we collected many more at this site.

We were at the Reckling Peak icefield, and a feature known to us as Elephant Nunatak[2] lay about 90 km to the west. This was our next destination, and as we traveled toward it I kept thinking about something that had happened a couple of years earlier.

Before the 1977 field season I had talked with Bill MacDonald, a photogeologist at the US Geological Survey, and Chief of their Branch of International Activities. He had supervised the US aerial photography program in Antarctica. This man had worked with thousands of photos of Antarctica. Becoming aware of our discoveries of meteorite concentrations on patches of exposed ice, he called my attention to a feature he called Elephant Nunatak, which was surrounded by extensive fields of exposed ice.

At the time we talked, he had learned that he had cancer and knew that he would die soon. I knew nothing of this, and some months later I decided to visit him for more detailed conversations about Elephant Nunatak and other possible sites. I called his office and a secretary told me he was in the hospital, in Annapolis, Maryland. She did not mention how grave his condition was, so I decided to drive down and pay him a visit. Late that evening, I left Pittsburgh for the five- or six-hour drive to Annapolis. I slept for an hour-and-a-half at a rest stop along the way. About 80 km outside of Annapolis, I drove into one of the most violent electrical storms I can remember, with high winds and rain so torrential that my windshield wipers could not keep pace. I had to pull over for about half an hour until it subsided.

It ended as quickly as it had begun. As I entered Annapolis shortly after dawn, the world appeared to be freshly washed, and glorious. The sky was a riot of shredded clouds and the recently risen sun turned them a bright gold in the east. I wished for a camera but had

---

[2] A nunatak is a small area of bedrock that has not been completely cloaked in ice – a rocky outcrop above the ice surface.

none, so I did what I could: locked the scene away in my memory; then found a restaurant and ordered breakfast. After eating, I located the hospital and inquired for Bill MacDonald at the Visitors' Desk. Checking in her files, the clerk at the desk said, "He's gone home." As I was turning away, in the same pleasant tone of voice, she said, "No, I'm wrong: he died last night."

In shock, I thought, "Maybe you phrased it better the first time," and wandered out. MacDonald Peak, near the north end of the main ridge of the Sentinel Range, in the Ellsworth Mountains, bears his name, but I shall always remember him for sending us to Elephant Nunatak.

The trip to Elephant Nunatak took a day, during which we coursed westward in a sort of swale 5–10 km wide between Griffin Nunatak to the north and, to the south, what appeared to be a swelling wave of ice – a frozen *tsunami* – that would shortly engulf us. We speculated later that this was in reality a downward step in the ice surface as it spilled over a subsurface ridge. This is typical of many of the places where we have found meteorite concentrations in the years since then. We also learned that "Elephant Nunatak" was not a nunatak, or rock outcrop, but only a moraine consisting of scattered rocks brought up from the bed of the ice sheet by action of the ice itself. So it became Elephant Moraine. We saw no need to change the "Elephant" part of the name, because its outline on aerial photographs (Figure 4.5) really is remarkably like that of an elephant, although with an oversize trunk. The "eye" of the elephant is the location that appeared to be solid rock, and led to its designation as a nunatak. This spot turned out to be just a concentration of moraine rocks dense enough to obscure the underlying ice. The Kyle party had made this same discovery the year before, but their report had not yet been published and we were unaware of it.

Elephant Moraine turned out to be a real bonanza of meteorites. Its patch of ice is one of a discontinuous series of exposed ice patches extending westward 90 km from Reckling Peak to Elephant Moraine and then somewhat northwestward beyond Elephant Moraine for

FIGURE 6.1 Lou Rancitelli and martian meteorite EETA 79001.
(Photo by W. Cassidy.)

20–30 km. All of the ice patches beyond Elephant Moraine had meteorite concentrations on their surfaces. In succeeding field seasons
we visited them all, giving them names such as Meteorite City, Texas
Bowl, Northern Ice Patch, Upper Meteorite City and Blue Lagoon.
The first morning, however, we were busy at Elephant Moraine, and
before lunch we had harvested a martian meteorite (EETA 79001)[3], as
shown in Figure 6.1.

Of course, in the field we didn't recognize it as a SNC meteorite,
but we did indicate in our field notes that it appeared to be unusual, so
it was the first one of that year's collection that was opened for examination when the samples reached Houston (see Figures 6.2 and 6.3).

[3] During the early years of the project it was anticipated that there might be more
than one field party engaged in collecting meteorites on the same patch of ice during
the same year. Meteorites collected during these years therefore had the letter "A"
added to their name in order to differentiate them from meteorites collected by
another field party that might bear the same number. Such meteorites would be
designated with a "B." This never happened, and eventually the convention was
dropped. Currently this specimen is more likely to be identified in the literature as
EET 79001.

FIGURE 6.2 EETA 79001 in the lab at Houston. The cube in the photo measures 1 cm along each edge. (NASA photo.)

Tables 6.1–6.5 list all the martian meteorites – antarctic and non-antarctic – that have been recognized as of the date of this writing.

MORE MARTIAN METEORITES

EETA 79001 was the first SNC meteorite that I remember collecting, but it was not the first we had collected. In 1979, the first report describing ALHA 77005 (found in 1977 at Allan Hills) had appeared in the literature. It was described as an igneous rock consisting predominantly of olivine, two species of pyroxene and maskelynite (shock-disordered plagioclase), along with a number of minor minerals. The fabric of the rock suggested that it was a cumulate – i.e., a specimen, some of whose constituent igneous minerals had settled toward the floor of the magma chamber, eventually being cemented together by solidification of the interstitial liquid between them. The composition

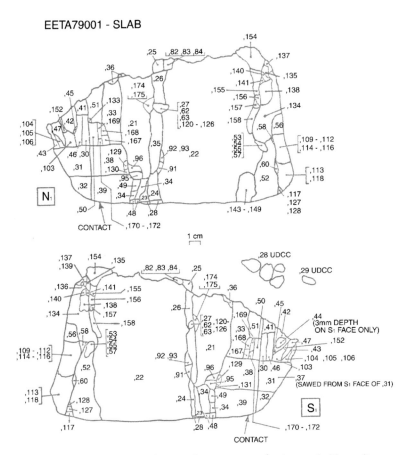

FIGURE 6.3 During the curation process at the Antarctic Meteorite
Laboratory at NASA's Johnson Space Center, it is usually necessary to
select subsamples for distribution to interested researchers. In the case
of martian meteorite EETA 79001, many, many requests were received.
The sampling procedure adopted by lab personnel was to cut a slab
from the specimen and then to physically break it into pieces, which
were then reassembled so that the relative position of each fragment
within the specimen could be recorded. Sawing, which was necessary to
produce the slab, can introduce traces of contamination from the wire
saw. Once the slab had been obtained, however, physically breaking it
apart into irregular-shaped pieces exposed fresh, uncontaminated
surfaces. Subsamples are designated by (comma number). The name of
subsample (,22) therefore is EETA 79001,22. Its position in the sample
relative to all other subsamples is accurately known from the diagram.

Table 6.1. *All known basaltic shergottites, as of Spring 2002. All specimens show strong similarities to Shergotty. They probably all originated as extrusive lava flows, but contain variable concentrations of minerals that had begun crystallizing before extrusion. Multiple specimens grouped together represent tentative fall-pairings. These are based only upon geographic proximity of the sites where they were found and general petrological similarity. Note that estimates of shock effects and crystallization ages are not available for many of the specimens.*

Meteorite name	Recovered mass (g)	Shock effects	Crystallization age (million years)
SHERGOTTY	5000	29 GPa	180 +/−20 to 190 +/−30
ZAGAMI	18000	29GPa	180 +/−20 to 190 +/−30
QUE 94201	12		327 +/−12
EETA 79001[a]	7942	34 GPa	180 +/−20 to 190 +/−30
DAR AL GANI 476[a]	2015		
DAR AL GANI 489[a]	2146		
DAR AL GANI 670[a]	1619		
DAR AL GANI 735[a]	588		
DAR AL GANI 876	6.216		
NWA 480[b]	28		
NWA 856[b]	320		
NWA 1068[b]	118		
NWA 1110[b]	576.77		
LOS ANGELES 1	452.6		
LOS ANGELES 2	245.4		
SAYH AL UHAYMIR 005	1344	strong	
SAYH AL UHAYMIR 008	8579		
SAYH AL UHAYMIR 051	436		

(*cont.*)

Table 6.1. (*cont.*)

Meteorite name	Recovered mass (g)	Shock effects	Crystallization age (million years)
SAYH AL UHAYMIR 094	223.3		
SAYH AL UHAYMIR 060	42.28		
SAYH AL UHAYMIR 090	94.84		
DHOFAR 019	1056		
DHOFAR 378	15		

[a] Contains lherzolite inclusions.

[b] NWA: Northwest Africa.

of ALHA 77005 could have been derived from the parental magma from which the shergottites also had formed or, alternatively, was a remnant of the source rock from which the parent magma of the shergottites had been derived by partial melting. So it was not an immediate member of the shergottite family of basalts; it was more like a cousin, or an aunt with a different family name: lherzolite.

EETA 79001, however, was brother and sister to Shergotty and Zagami. Incredibly, it was two rocks in one: a relatively finer-grained igneous rock (lithology A) whose crystal size was smaller than that of Zagami, and a coarser-grained igneous rock (lithology B) whose crystal size was smaller than that of Shergotty but larger than that of Zagami. There is a line of contact between the two lithologies, and they are considered to represent successive magmatic pulses. It is as if a basaltic lava flow, containing some crystals that had already formed in the magma chamber by partial crystallization, flowed out on the surface where it continued to crystallize until it was solid rock. Some time later a second lava flow was emplaced above the first. The resulting rocks had different grain sizes, either because their crystallization histories had progressed to different stages of completion at the time that each was extruded or because one of them was a thicker

Table 6.2. *All known lherzolitic shergottites, as of Spring 2002. All specimens show relationships to Shergotty, but are probably plutonic cumulates in origin. In contrast to the basaltic shergottites, these may have formed by accumulation of crystals on the floor of a magma chamber, and were never extruded at the surface. None of these specimens appear to be fall-paired. Estimates of shock effects and crystallization ages are available for only one specimen.*

Meteorite name	Recovered mass (g)	Shock effects	Crystallization age (million years)
ALHA 77005	482	43 GPa	187 +/−12 to 154 +/−6
Y 793605[a]	18		
LEW 88516	13		
GRV 99027[b]	9.97		

[a] Y: Yamato Mountains.

[b] GRV: Graves Nunataks.

flow that cooled more slowly than the other after emplacement. It is this difference in grain size that marks the contact between the two rocks.

Interestingly, lithology A had some foreign inclusions that could have been torn off the walls of a magma chamber, or a volcanic neck, and incorporated into the magma. These inclusions are fragments of lherzolite (like ALHA 77005), so there are valid reasons to suspect relationships between EETA 79001, ALHA 77005, Shergotty and Zagami. Do these relationships indicate a common origin in the same magma chamber for all the basalts and lherzolites, or nearby origins in separate igneous environments? Predictably, opinions differ on the answer to this question.

EETA 79001 was an important find for another reason: it contained proof that SNC meteorites come from Mars. This meteorite had been highly shocked during the impact that ejected it from the martian

Table 6.3. *All known nakhlites, as of Spring 2002. All specimens are clinopyroxenites and show strong similarities to Nakhla. They probably all originated as shallow intrusions. Multiple specimens grouped together represent tentative fall-pairings. Note that estimates of shock effects and ages of crystallization are not available for all specimens.*

Meteorite name	Recovered mass (g)	Shock effects	Crystallization age (billion years[a])
NAKHLA	9000	v. little	1.3
LAFAYETTE	800	v. little	1.3
GOVERNADOR VALADARES	158	v. little	1.3
Y 000593[b]	13713		
Y 000749[b]	1283		
NWA 817[c]	104		

[a] billion: thousand million.

[b] Y: Yamato Mountains.

[c] NWA: Northwest Africa.

Table 6.4. *There is only one chassignite known. It is a dunite (consists principally of olivine). It probably formed in a shallow intrusion.*

Meteorite name	Recovered mass (g)	Shock effects	Crystallization age (billion years[a])
CHASSIGNY	4000	15–35 GPa	1.3

[a] billion: thousand million.

surface, and one of the shock effects was to produce pods of glass that had been melted from the constituent minerals of the meteorite by the transient heating generated by shock pressures. The melts that were formed were partly injected along cracks in the rock and partly

Table 6.5. *There is only one martian orthopyroxenite known. It consists principally of the mineral orthopyroxene and probably formed as a plutonic cumulate, i.e., by accumulation of orthopyroxene crystals on the floor of a magma chamber. Note that its estimated age is more than three times as great as any other known martian meteorite, and therefore it may have formed less than 100 million years after the planet itself.*

Meteorite name	Recovered mass (g)	Shock effects	Crystallization age (billion years[a])
ALH 84001	1931	intense, multiple	4.5 +/−0.13 GA

[a] billion: thousand million.

retained as molten beads at the sites where they formed. Cooling occurred immediately behind the shock wave and the beads were chilled to glass before they could crystallize. Apparently the crater-forming asteroid had built up a lens of compressed atmosphere in front of it during its lengthy trajectory toward the surface of the planet. When it struck the surface it injected highly compressed air into the target rock, and some of this was trapped in the shock-melted inclusions. We know the composition of the martian atmosphere from measurements made by Viking Landers I and II. When some of the glass inclusions were picked out of EETA 79001 and remelted, they gave up their dissolved gases. These gases, when analyzed and corrected for slight terrestrial contamination, contained nitrogen and carbon dioxide in the same abundances as the atmosphere of Mars; they also had isotopes of argon, neon, krypton and xenon in the same abundances as does the martian atmosphere. This neat bit of detective work by a number of workers for the first time tied a SNC meteorite directly to the planet Mars and, through this meteorite, to all the other SNC meteorites.

The nakhlites and Chassigny are very similar to each other and display preferred crystal orientations suggestive of crystal settling within a magma body, but the inferred source rocks of the magma from which these igneous meteorites formed were somewhat different in composition than the source rocks of the magma from which the shergottites crystallized. In any case, it is easy to believe that the martian meteorites we had collected through 1979 came from at least two igneous bodies of somewhat different composition, at or near the surface of Mars.

In the 1984 season, we found our third martian antarctic meteorite, ALH 84001. This meteorite is very similar in its macroscopic appearance and mineralogy to a group of non-martian, asteroidal meteorites called diogenites, and this one initially was misidentified as a diogenite. ALH 84001 was found on the Allan Hills Far Western Icefield by the 1984–85 field party; specifically by Roberta Score, who at that time, aside from being an ANSMET team member, was Secretary of the Meteorite Working Group and Lab Manager of the Antarctic Meteorite Laboratory at the Johnson Space Center.

The Far Western Icefield is an irregular patch of exposed ice (see Frontispiece) where we first landed by mistake, in 1982. We had planned to camp at the Allan Hills Middle Western Icefield, but it had been a somewhat hazy day and our helicopter pilot had circled around, out of sight of visual landmarks, looking for the icefield that we had spotted on satellite photos. We strayed farther and farther out over the ice plateau, and finally spotted the ice patch we later called the Allan Hills Far Western Icefield. Thinking it was the Middle Western Icefield, we set down and hurriedly unloaded our equipment and supplies; fuel was low and our pilot was getting nervous. We were not successful in making radio contact with McMurdo, and the rules dictated that we had to make contact immediately or we could not stay. We left our cargo on the ice and lifted off, still convinced we had been on the Middle Western Icefield, but from our return track the pilot later figured out exactly where we had been.

On the return flight we discovered that the crewman had picked up a meteorite next to our landing spot; this was a good sign. A bad sign was given by the fuel gauge, which was alarmingly low. A conference between the pilot, copilot and crewman was followed by a somewhat tense conversation between the crewman and us, in which he reminded us of the routine for emergency evacuation of the helo, in case we ran out of travel before the end of the trip. Essentially, we would autorotate to a gentle landing (the optimistic viewpoint) while curled into the fetal position around our seat belts (the pessimistic provision). When we had set down, the crewman would direct us out that side of the helicopter that gave us the best chance of evading the possibly still rotating propellor blades and, erupting from the cabin like beans from a bin (actually, his simile was noticeably more graphic, involving one of the ends of a cow), we would do the 100 m at our best speed, in case the aircraft caught fire. Luckily, this did not become necessary. We made it to our refueling stop at Marble Point, landing on fumes, and later found we had discovered, by mistake, a meteorite concentration at the Allan Hills Far Western Icefield.

LEW 88516, a very small, 13.2 g martian meteorite, discovered on the Lewis Cliff Ice Tongue on December 22, 1988, is very similar to ALHA 77005. There are a few minor petrographic differences between them but these differences are so small that both rocks could have crystallized in the same magma chamber. Compositionally, they are both related to the foreign inclusions in EETA 79001, and through EETA 79001 to Shergotty and Zagami.

QUE 94201 is a very small, 12 g martian meteorite, discovered by the 1994 field party on an ice patch around the southern end of the Queen Alexandra Range, where this range of mountains is submerged in the plateau ice of East Antarctica. It appears to be a basalt in composition. Some of the major minerals in this rock show extreme compositional zoning, in which the core of a mineral grain is different compositionally from its outer layers. This is characteristic of continuous crystallization in a magma cooling too fast for chemical readjustments to occur within the already formed center of the crystal. QUE 94201 appears to be similar to parts of Zagami, and thus is another shergottite.

Y 793605 is a 16 g martian meteorite discovered in the Japanese collection from the Yamato Mountains Icefield. This specimen has been brecciated and the plagioclase has been shock-disordered, but it was originally a cumulate having large crystals 1–7 mm across. It appears to have affinities to both ALHA 77005 and LEW 88516, so it is another lherzolite.

COLD DESERTS...HOT DESERTS

There are two reasons why concentrations of meteorites can occur in the cold desert that is Antarctica: the cold temperatures and lack of exposure to liquid water greatly inhibit the process of chemical weathering, so meteorites can be preserved for great lengths of time; and the slow movement of the ice sheet toward the coastline sometimes ends prematurely at a surface where ice wells up against a barrier and disappears by ablation. In such an area the arriving ice disgorges any meteorites it has collected during its travels and leaves them stranded on the ablation surface. Because they do not weather very fast even when exposed to the cold, dry air, meteorites can become concentrated on ablation surfaces if the local conditions leading to ablation have persisted long enough. So in Antarctica, the process of meteorite concentration can occur both across a distance, represented by the upstream path length of the ice, and also through time, made possible by the slow weathering rates.

As these insights became general, collectors began turning their attention to the hot deserts of the world, reasoning that even though meteorites would not be expected to be transported horizontally as in antarctic ice, they could be preserved over great lengths of time because hot deserts also are very dry. An added advantage of hot deserts is that, while most of them occur in remote corners of the world, they are still more accessible than is any part of Antarctica, and meteorite-searching expeditions can be much less expensive to mount. This reasoning has paid off, mainly in a spate of meteorite recoveries in parts of the Sahara Desert located in Libya and Algeria, on the Arabian Peninsula and on the Nullarbor Plain in Australia. Notice in Table 6.1 that a number of basaltic shergottites from hot deserts are listed. These are the fall-paired Dar al Gani, Sayh al Uhaymir, Dhofar and Los Angeles specimens. Dar al Gani is a location in the Libyan Sahara and Sayh al Uhaymir and Dhofar are on the Arabian Peninsula. Some might consider Los Angeles to be a *cultural* desert, but the days when it were a small town *in* the desert are long gone. The Los Angeles pair was picked up some years ago by a rockhound in the Mohave Desert of California, but he recently *rediscovered* them in his garage in Los Angeles, thus the misleading designation of Los Angeles for these paired specimens.

THE QUESTION OF INTERPRETATION

Study of the martian meteorites gives us an ever-expanding store of facts about the variety of igneous rocks we can expect to find when we finally visit Mars. From a knowledge of the igneous rock types present, we can travel a step further back to speculate on the compositions of the parent magmas that gave rise to these meteorites by different degrees of partial melting. In each step back, of course, our conclusions become less certain, but each additional martian meteorite we find has the potential to strengthen or change these conclusions because of the information it carries about its origin.

What can we say about the conditions of origin of the martian meteorites? Certainly, they are all igneous rocks, similar in composition either to terrestrial basalts or to the rocks that can form by fractional crystallization of a basaltic magma, followed by accumulation of the early-formed crystals. The igneous histories of basaltic shergottites, at least as far as we can decipher them, do not seem to involve quiescent processes in which a constant bulk composition in a closed system crystallizes over a temperature interval to its constituent minerals. Instead, the petrologic histories of most of these rocks seem to have involved: (1) admixtures of liquids, crystals or even rock fragments that are out of equilibrium with the crystallization stage that had been reached; and (2) a final stage of more rapid cooling than had occurred earlier. Such a history speaks of partial crystallization during turbulence and mixing in a magma chamber while forces are building toward an eruption. Eventually an eruption occurs, as the slurry of crystals and still-molten liquid is extruded in thick flows, within which faster cooling occurs.

The fall-pairings tentatively suggested in Tables 6.1 and 6.3 are based on the fact that the suggested pairs are petrologically similar and were found in relatively close proximity. If these fall-pairings are correct, the world's collection contains 20 martian meteorite falls, as of the date of this writing. Eight are from Antarctica and 12 were found elsewhere. Martian meteorites are still falling: four of the non-antarctic falls are observed falls. Six of the eight we have found in Antarctica are shergottites.

The composition of ALH 84001 shows some similarities to basalts and lherzolites, but isn't quite one of them. It also shows some similarities to nakhlites and Chassigny, but is not quite one of them, either. This meteorite from the Allan Hills Far Western Icefield was originally classified as a diogenite, one of the asteroidal igneous meteorite groups. Even though its recognition as martian is only recent, this extremely interesting bit of Mars has become perhaps the most intensively studied of all known meteorites. The age of formation of this igneous rock is not the 1.3 billion years, or even less, that we have learned to expect of martian meteorites – it is about 4.5 billion years. Thus, ALH 84001 crystallized only about 100 million years after its parent planet formed, so it must have come from a location on Mars having many more ancient rocks than all the others. This meteorite may provide a key to unlock some of the unknowns about the early days of Mars. Even though its age of formation is 4.5 billion years, in common with all other *asteroidal* igneous meteorites, we do not class it with them. Why? – because its oxygen isotopic $^{17}O/^{16}O$ and $^{18}O/^{16}O$ ratios identify it as martian.

HOW MANY IMPACT SITES HAVE THE MARTIAN METEORITES COME FROM?

How many asteroidal impacts (and thus, how many impact sites on Mars) were required to produce the meteorites we have already collected? Based on petrological similarities, there may have been only three (i.e., basalts/lherzolites, clinopyroxenites/dunite, and ortho-pyroxenite), or conceivably there may have been as many as 20 – one for each martian meteorite fall. Otto Eugster and his colleagues at the Physical Institute, University of Berne, Switzerland have measured the lengths of time that several of the martian meteorites have been exposed to cosmic radiation in space. Summarizing these "cosmic-ray ages" and those of other martian meteorites determined by other workers, they suggest that it may have required five or six separate impact events on Mars, occurring over the last 15–20 million years, to produce projectiles from Mars that have fallen to Earth as meteorites. A number of martian meteorites have been found since the measurements of Eugster and his colleagues, but most of these do not have the data necessary to update Eugster's estimate.

WHERE ON MARS ARE THE IMPACT SITES?

If we knew the various locations on the surface of Mars from which these meteorites came, we would have taken a major first step toward mapping the igneous geology of the planet. Mars is pockmarked with impact craters, but only a limited number of them are big enough to have had the requisite energies of formation. Many observers also still favor the idea that material can be thrown off Mars more easily by an oblique impact, and only a fraction of the largest craters are elongated enough to qualify as such a site. Current betting is that ALH 84001, with its ancient formation age, came from the most ancient terrain on Mars – the densely cratered highlands in Mars' southern hemisphere. But the probable age of the impact that separated ALH 84001 from Mars is only about 16 million years, so the crater itself must be a fresh-looking one. Nadine Barlow, at the University of Central Florida, has nominated an elongated crater in the Sinus Sabaeus region about 14° south of the martian equator as the most likely candidate.

The other martian meteorites have much younger formation ages and must have come from younger terrains. The most recent younger terrains on Mars are extrusive flows from large-scale volcanism in the Tharsis region, where a group of tremendous shield volcanoes is located. Most of these are as large or larger than any shield volcanoes on earth. Mauna Loa, for example, earth's largest volcano, rises about 10 km above its base on the sea floor. Olympus Mons, the largest shield volcano on Mars, rises 26 km above the surrounding plain! Ceranius Tholus, another member of this group, is about the same size as Mauna Loa, and has a large, elongated impact crater impinging on its lower flank. This crater-forming impact must have sampled the igneous rocks of which the tremendous pile of Ceranius Tholus is built, and its pronounced elongation suggests a very low-angle impact, relative to the surface. These features nominate it as an excellent candidate to have supplied rocks thrown off Mars by asteroidal impact. Perhaps one or more of these rocks are now in our antarctic meteorite collections...

MARS AND COMPARATIVE PLANETOLOGY
Those who study meteorites tend to be a sentimental lot – we love to get together and talk about the "Old Days": the origin of the earth, the origin of the solar system and its evolution from a primordial cloud, and things like that.

We have studied terrestrial geology for a couple of hundred years now, and our state of knowledge about the earth's surface and inferences about its interior could be compared to the state of knowledge of an anatomy student who has a very complete understanding of the structural details of, say, the pig, but he can only generalize his findings to all vertebrates in a very limited way. When he has performed the same research on many vertebrate genera, he has a much better feel for the ways in which one genus may differ from another, and the ranges of differences that are possible. His study is called comparative anatomy. Geologists, with their growing understanding of the earth, have been in the position of the anatomy student after learning about his first genus, but samples from other planets are necessary before we can reach any generalizations that could be called comparative planetology. We made it our business to obtain lunar samples, so we could begin to compare the earth and its moon, and remote sensing studies are supplying much information about other solar system bodies. So far, however, meteorites are the only actual samples of bodies other than the earth and moon that we can examine in the laboratory. Meteorites represent many different parent bodies, but, with the possible exception of one group (the HED meteorites, which *may* have originated as parts of Asteroid 4 Vesta) we know very little of where in the solar system each parent body was located and what its overall compositions and structures were. By great good chance we have martian meteorites, which can begin to tell us something about that planet, and lunar meteorites which can supplement what we already knew about the moon from samples that have been collected there. So we can say that we have embarked on a new discipline: comparative planetology. The study of meteorites makes vital contributions to that field of knowledge.

The role of martian meteorites in comparative planetology
Even with all we know about the geology of the earth, our understanding of the mantle and core of our own planet still has speculative aspects. These uncertainties provide seemingly irresistible opportunities to propose not-impossible schemes for the way the earth formed, and by extension the way or ways in which all the planets formed.

Assumptions often made for the formation of the earth range between: (1) accretion at low temperature, slowly enough so that heat generated by gravitational energy of infall and compaction could be radiated away, resulting in a homogeneous body that later heated by decay of radioactive species and fractionated into core and mantle; and (2) rapid accumulation at temperatures that were high enough so that fractionation occurred during condensation, leading to accretion of refractory materials first, to form the earth's core, followed by mantle formation from less refractory materials under cooler temperatures. More or less between these extremes is the idea that the earth started to accumulate cold and formed a nucleus of volatile-rich material as an initial core. As it grew larger, the gravitational energy of accumulating material was released as heat. This caused successive outer zones to become hotter and hotter until iron oxide (FeO) was reduced to the metallic state and a tremendously dense atmosphere made up of volatiles and near-volatiles enveloped the forming earth. When a large amount of liquid metallic iron had collected near the surface, it penetrated to the core of the planet in a massive overturning process that displaced the primitive, volatile-rich nucleus upward to mix with the mantle material.

This very rapid summary has omitted all the detailed arguments and discussions of consequences that embellish all of these ideas; furthermore, I have not provided citations for any of them, since I hoped only to indicate the wide variety in versions of planetary formation that have been offered over the years. The point is, that being the only samples we have of Mars, martian meteorites furnish a suite of rocks whose point of origin we know relative to the earth's distance from the sun; they are from a planet that grew only to about 10% of the mass of the earth, and whose geology is different in many ways from that of the earth. We can now speculate about the "Old Days on Mars;" this, in combination with our speculations about the early days of Earth, can help us to understand planetary formation.

THE GEOCHEMISTRY OF MARS

Geochemists understand that in igneous, metamorphic and sedimentary processes groups of elements tend to act in similar ways, depending upon their degrees of ionization and their effective sizes. For example, a large group of elements "prefers" to form oxides, of which the largest group, by far, is the silicates. Silicate minerals form the predominant rocks in the crust and mantle of the earth. Elements that like to form oxides are called *lithophiles*. Another group of elements prefers an alliance with sulfur and sulfur-like elements to form sulfides, and these are called *chalcophiles*. A third group seems to be most at home, particularly at high temperatures, in the metallic state. They have chemistries similar to iron and are called *siderophiles*. These are just generalities, however – in the presence of an excess of oxygen, most siderophiles will oxidize; similarly, some chalcophiles can become siderophiles if both sulfur and oxygen are scarce; or they can become oxides if sulfur is scarce but oxygen is abundant. We may find the element calcium, for example, present in both sulfides and silicates. Another example is iron, which undoubtedly is the principal element in the metallic core of the earth, but is also a major oxidized component in the principal silicate minerals of the mantle and crust of the earth. Incidentally, iron is also common in sulfides. So the composition and structure of a differentiated planet depends on a large number of variables, including the relative abundances of all the elements that form the planet and the temperatures and pressures to which they have been subjected.

Elements and compounds can be characterized as volatiles or refractories. Some elements and compounds tend to vaporize at lower temperatures than others; these are described as *volatiles*. Conversely, some elements and compounds tend to resist vaporization until very high temperatures are attained. These components would be the first to condense from the vapor state, even at very high temperatures; they are called *refractories*. In between is a range of elements and compounds that vaporize more or less readily relative to each other.

In considering hypotheses for the formation of a planet, the geochemist must consider whether the body aggregated at a high or low temperature, what the relative abundances of the elements and molecules might have been in the source material (i.e., the solar disc and any solid bodies that had already formed within it), what the vapor pressure might have been in the solar disc because this will influence condensation histories, and, under these conditions, which elements will be siderophile, chalcophile, or lithophile. Then he or she must choose the formation history that seems to conform most closely to what is observed and inferred in the planet itself. Because of the number of variables involved in these studies, one could conclude that geochemists will always have work – and this is true.

WHAT DO THE MARTIAN METEORITES TELL US ABOUT MARS?

The martian meteorites are so far the only samples we have of that planet. As such, they are our only source of information on Mars that can be examined and analyzed in the laboratory; so they provide a nucleus of data that could be used in planning our future surface investigations, as well as in beginning to understand the geology of Mars. Until we can directly sample the martian surface, much of our speculation about the origin and history of Mars will continue to be influenced by what we have learned from these priceless specimens. Answers to the following questions indicate some of the ways in which martian meteorites have influenced our thinking about Mars.

When did Mars differentiate into core and mantle, and what do we know about its subsequent petrogenetic history?
The martian meteorites contain residual amounts of certain radioactive isotopes, namely, potassium-40 (^{40}K), rubidium-87 (^{87}Rb), uranium-238 (^{238}U), uranium-235 (^{235}U) and thorium-232 (^{232}Th). These isotopes are left over from the time at which their parent planet underwent a major differentiation into core, mantle and (perhaps) crust. Each of these radioactive species has a very long half-life, as shown in Table 6.6.

The significance of such long half-lives is that measurable amounts of the original, or parent, isotopes are still present. At the same time, significant amounts of the ultimate daughter products formed by their radioactive decay processes have accumulated and

Table 6.6. *Parent radioactive isotopes that generate heat by their decay; half-lives of decay; and stable daughter products of each*

Parent isotope	^{40}K	^{87}Rb	^{238}U	^{235}U	^{232}Th
Half-life (billion years)	1.25	48.8	4.468	0.7038	14.01
Daughter isotope	^{40}Ar	^{87}Sr	^{206}Pb	^{207}Pb	^{208}Pb

The daughter products are argon-40 (^{40}Ar), strontium-87 (^{87}Sr), lead-206 (^{206}Pb), lead-207 (^{207}Pb), and lead-208 (^{208}Pb). Decay of ^{40}K (potassium-40) produces ^{40}Ca (calcium-40) in a constant, known ratio to ^{40}Ar, but the ^{40}Ca is not useful in age determinations because large and variable amounts of non-radiogenic ^{40}Ca already exist in most igneous rocks at their times of formation.

Billion years = thousand million.

their relative concentrations also can be measured. ^{40}K is less reliable, however, because its daughter product, ^{40}Ar, is a volatile and may have been depleted on subsequent heating.

We now know from whole-rock ages of martian meteorites that Mars differentiated into core and mantle during or shortly after formation. This means that it was hot during planet formation, or else heated up shortly after the planet formed. Two types of measurements can be made: (1) The parents/daughters ratios can be measured for the whole rock; and (2) ratios of parents/daughters can be determined for individual minerals that make up the whole rock. While these ratios for the *whole rock* are believed to determine how long it has been since the parent planet differentiated into the major divisions of core, mantle and crust, later metamorphic heating or complete melting (igneous) processes could have redistributed the parents and daughters between different minerals within the whole rock, so that the clock for the individual minerals would then be reset. Thus, while the whole rock ratios would not have changed, and should still give us the date of primary fractionation of the planet, the ratios within individual minerals could have changed. Since the clock had been reset for the minerals, the ratios of parents/daughters in the minerals in a given rock, measured today, would determine how long it had been since the last metamorphic or igneous episode. All of this depends on the assumption that nothing has been added to, or subtracted from, the material making up the whole rock since the original planetary differentiation; even during whatever metamorphic and/or igneous processes have affected it. We can gain confidence in the whole-rock age determinations because all the martian meteorites, even though they may have come from several different localities on the planet, agree in giving times of planet-wide differentiation close to −4.5 billion years. This is only 100 million years, or so, after their parent planet formed. Other (non-martian) igneous meteorites also give whole-rock ages around 4.5 billion years, and this forms the basis for our conclusion that all large solid bodies in the solar system underwent early heating that was intense enough to cause planetary differentiation into core, mantle, and possibly crust, at roughly the same time.

Since the martian meteorites are all igneous rocks, and the individual mineral ages of different ones were reset at a variety of times, we can suspect that Mars has had episodes of volcanic activity during its entire history. Before lunar samples and martian meteorites, we were able to say that the parent bodies of all known non-martian igneous meteorites had no history of igneous processes much beyond the original differentiation at 4.5 billion years. The reason is that their *individual-mineral times of formation* were also −4.5 billion years. The surface of Mars, however, shows evidence of volcanic activity during the relatively recent past, and while whole-rock ages give the original differentiation of Mars into core, mantle and crust as −4.5 billion years, ages determined from individual-mineral isotopic measurements suggest that the basalts and one lherzolite (ALHA 77005) crystallized 180–190 million years ago, while clinopyroxenites and the one dunite (Chassigny) crystallized 1.3 billion years ago. As noted above, one of the antarctic martian meteorites, ALHA 84001, has been determined (based on somewhat uncertain measurements) to have crystallized much earlier than these dates, and may in fact represent original, ancient martian crust.

Is Mars really very "dry?"

Martian meteorites contain the highly volatile elements fluorine, chlorine and bromine in much greater abundance than do terrestrial rocks. This suggests that that part of the solar nebula from which Mars formed had much higher abundances of volatiles than the region in which Earth formed. A highly volatile compound, water, should also be present in great abundance in such a source region. But while martian meteorites are rich in other volatiles they are remarkably poor in water, as measured by their almost complete lack of hydrated minerals. Similar terrestrial basaltic rocks often contain much higher proportions of hydrated minerals. The conclusion from this is that while Mars initially must have accumulated a significantly higher proportion of water than the earth, Mars' mantle somehow lost a much higher proportion of its water, while at the same time not losing its other volatiles.

There is other evidence for a wetter past. Major parts of the martian surface are dissected by water courses, now dry. Some workers claim evidence that Mars even had a shallow ocean in its northern hemisphere, complete with gouges on the bottom made by icebergs drifting southward from the polar ice cap. Some water remains as ice, trapped in the polar caps, and there may be substantial quantities existing as permafrost-like ice deposits below the surface. Judging from the abundances of other volatiles, however, Mars started out with enough water to cover its entire surface to an estimated depth of 130 m. How to explain that very little of Mars's original water is left today?

Gerlind Dreibus and Heinrich Wänke, of the Max Planck-Institut für Chemie, Germany, note that oxidized iron is more abundant in the martian meteorites than in terrestrial rocks. They infer that a very large proportion of the initial water accumulated by Mars reacted in the mantle with metallic iron, trapping oxygen as iron oxide. The hydrogen that was liberated as a result then escaped the planet. When all the iron in the mantle had been oxidized in this manner, only a small inventory of water remained, leaving behind a very dry planet that had started out wetter than the earth.

What is the thermal history of Mars?

Apparently Mars has had igneous activity during most of its history. As discussed earlier, because whole-rock ages for the martian meteorites agree at 4.5 billion years we believe that Mars was hot after accumulation had occurred and that it differentiated at least into core and mantle fractions soon after the planet formed. But magmas from which these meteorites crystallized, some of which apparently existed only as recently as 1.3 billion years and 180 million years ago, cannot have been produced by residual heat of accumulation, because such heat would have been dissipated early in martian history.

The agents that keep planets warm over the ages are long-lived radioactive species that release heat during decay. Martian meteorites have potassium/uranium ratios comparable to those of Earth, but their thorium/uranium ratio is significantly lower, so of the three planetary heat-producing nuclides, Mars appears to have less thorium, relative to the earth. Even with this smaller contribution from thorium, however, Mars has retained enough heat to generate magma very recently. On Earth, an extreme compositional differentiation has occurred between the mantle and crust and, because of their geochemical characteristics, most of the heat-producing species have been fractionated into the granitic crust, from which heat is lost more readily. On Mars, a less extreme mantle–crust differentiation may have occurred. This would make it possible for higher concentrations of potassium, thorium and uranium to have remained in the mantle, at depths from which it would be less easy for heat to escape without melting rocks. We can conclude that, unlike the earth, with its highly differentiated granitic crust, there probably are no granitic-type rocks at the martian surface.

Do we have any clue as to the parent magma(s)
of SNC meteorites?

The terrestrial rocks that most closely resemble martian meteorites are basalts. There could have been at least three parent magmas for the known SNC meteorites, but all three had two major differences from terrestrial basalts. Relative to terrestrial basalts they were all sodium-rich and iron-rich. Beyond this conclusion, speculations are based on our understanding of the crystallization process for igneous rocks.

When magmas crystallize, they do so not at one fixed temperature, but over a range of temperatures. Most rock-forming silicate minerals can vary in composition, in response to temperature. Usually this variation, which we call solid solution, is expressed as differences in magnesium/iron (Mg/Fe) or calcium/sodium (Ca/Na) ratios. The general pattern is for the highest-temperature mineral to begin crystallizing first, and then continue to crystallize as the temperature drops, usually changing composition somewhat in response to that Mg/Fe or Ca/Na ratio that is most comfortable being in that mineral structure at that temperature. High Mg/Fe and Ca/Na ratios are typical of the high-temperature ends of the petrologically important solid solution series, and as the temperature drops the minerals become richer in Fe and Na, respectively, while becoming poorer in Mg and Ca.

Except under special circumstances, the *composition* of a magma that is completely molten will dictate what mineral will begin to crystallize first as it cools down, while the *temperature* will dictate what Mg/Fe or Ca/Na ratio the first-crystallizing mineral will have. As the first mineral continues to crystallize and change composition with falling temperature, second, third, and fourth mineral phases can appear and crystallize simultaneously with the earlier phases. Often, the higher-temperature minerals will finish crystallizing while there is still liquid remaining and the remaining liquid then will contribute only to the lower-temperature minerals that are still forming. Eventually, no liquid remains and the resulting interlocking crystalline assemblage is solid rock, which continues to cool as heat is dissipated. Under the conditions just described, the bulk composition of the rock is the same as the composition of the magma from which it formed, and chemical analysis of a representative sample of that rock will specify the composition of the parent magma.

Nothing in Nature is ever perfect, however, and the general situation in a magma chamber is that the early-formed crystals will separate themselves from the parent magma, usually by sinking (but see the following chapter for a major exception, plagioclase, to the sinking tendency). Chemical reactions in silicate solutions are sluggish, and so as these early-formed crystals pile up on the floor of the magma chamber they have effectively removed themselves from further reactions with the bulk of the remaining liquid. This process is called fractionation, and it helps to account for the wide variety of igneous rock compositions that we see. Because the high-temperature minerals have selectively removed elements of which they are composed from the parent magma, the remaining magma has a different composition than the original parent magma. Rocks subsequently formed from the accumulated crystals and their intercrystalline liquids have a different composition from the parent magma and are termed "cumulates." If the overlying, evolved liquid is then extruded onto the surface as a lava flow, the bulk composition of the resulting extrusive rock will be the same as that of the remaining liquid, and not of the original parent magma. Sometimes, also, volcanic eruptions are so violent that rock fragments can be torn off the walls of the volcanic neck and incorporated into the erupting magma, and this can further change the bulk composition of the resulting rock.

One should not expect to be an expert petrologist from the foregoing summary, but at least one can begin to understand the potential problems that can be encountered in trying to estimate the composition of the parent magma from which a given rock formed. These estimations can involve guesses about the proportion of original cumulate minerals in a given meteorite, and then correcting the bulk composition to allow for overabundances due to crystal settling. Another method, among the many that have been tried, is to find tiny melt inclusions in some of the first-formed high-temperature minerals, determine the compositions of these melt inclusions and assume they represent samples of the original parent magma, trapped within another crystal at the onset of crystallization.

So we see that estimates can be made in a number of different ways to define compositions of the parent magmas of igneous rocks. For the martian meteorites these estimates tend to cluster within three discrete composition ranges. The parent magmas that seem most likely would have produced the basalts/lherzolites, the orthopyroxenite, and the clinopyroxenites/dunite. While the clinopyroxenite/dunite and orthopyroxenite composition ranges are close enough to suggest a possible relationship between their parent magmas, the basalt/lherzolite range is different enough from the other two to suggest that it had a different parent magma.

How has the atmosphere on Mars evolved?
Water, carbon dioxide and nitrogen were present in greater concentrations in the martian atmosphere early in its history, and the early martian atmosphere was denser, possibly much denser, than it is today. How do we know this?

We know from observations on the earth, as well as from thermodynamic calculations, that whenever a phase change occurs, such as melting of water ice to liquid or vaporization of liquid water, heavier isotopes of the same element tend to remain in the lower-temperature phase. Also during the reverse process, condensation of vapor to liquid or crystallization of liquid to solid, heavier isotopes tend to enter the lower-temperature form preferentially. This is a very subtle effect, but over millions of years, if very little mixing occurs between atmosphere and hydrosphere and hydrosphere and crust, water with a higher proportion of heavier isotopes can be found in the crust, and water with a higher proportion of lighter isotopes would be found in the atmosphere. Perhaps a more important effect, however, is that at the outer fringes of any planetary atmosphere, the relatively lighter atoms and molecules are being lost to space, preferentially.

Oxygen, carbon, nitrogen and hydrogen in martian meteorites are isotopically much heavier than in both terrestrial rocks and the most primitive asteroidal meteorites. Because this was true as late as 1.3 billion years ago on Mars, when some of the martian meteorites crystallized, we can assume that major fractionations by degassing had already occurred earlier in martian chronology, and the early Mars atmosphere was much richer in these elements.

Is there, or has there been, life on Mars?
Many have speculated. See the following extract, followed by an imaginary interview with me.

> She was as destitute of clothes as the green Martians who accompanied her; indeed, save for her highly wrought ornaments she was entirely naked, nor could any apparel have enhanced the beauty of her perfect and symmetrical figure. [Edgar Rice Burroughs, *A Princess of Mars*]

Q.   *You say you read this as a boy?*
A.   *Yes.*

Q.   *Did your mother know you were reading these stories?*
A.   *No.*

Q.   *Do you believe there is life on Mars?*
A.   *I would sure like to.*

Our most ancient martian meteorite, ALH 84001, has become a subject of intense controversy. A group led by David McKay at the Johnson Space Center and some of their colleagues at other institutions recently described extremely tiny structures inside ALH 84001 that to them suggested fossilized organisms. Associated with these forms were accumulations of polycyclic aromatic hydrocarbons (PAHs: a group of organic compounds that can be produced by the decay of formerly living creatures) and tiny crystals of magnetite and iron sulfide that could be products of chemical processes that occur in some terrestrial organisms. All of these were embedded in, or were preferentially associated with, tiny carbonate globules or "rosettes" whose precursor solutions might have provided a life-supporting environment. They suggested that all of these findings, combined, reinforce each other to support a hypothesis that living organisms were present early in the history of Mars. These are exciting thoughts indeed, because they suggest that life might arise independently almost anywhere, if given only the minimum of environmental encouragement.

There may be a loneliness in our collective psyche that can only be satisfied if we find we are not the only intelligent beings. Mankind has its dreams, and one of them is that one day we will meet living beings from places other than Earth, with whom we will be able to converse, exchanging ideas and congratulating each other for attaining the intellectual level necessary to begin to comprehend the universe. Of course, there is an unspoken corollary to this dream: the others should not be vastly superior to us!

At any rate, if the McKay group's findings were to be supported by additional evidence, ALH 84001 could perhaps become, in human experience, the most important rock specimen ever collected.

But one must pay attention to the critics, and there are many. It cannot be shown at this time whether or not these tiny structures really are fossilized life forms. The structures themselves do have fascinating shapes, but how do we know they are not just artifacts of the examination process, produced commonly at this scale as undulations in the fractured surface of a mineral, or as lumps produced in the process of coating the surface with vapor-deposited gold, as is done routinely with transmission electron microscope samples such as these? Also, PAHs are not exclusive products of living forms – they are everywhere and even can be collected from automobile exhaust gases[4]. Magnetite and iron sulfides can be, and usually are, products of inorganic reactions. Finally, the culture medium itself – the solutions that gave rise to the carbonate globules – might actually have formed at temperatures so high that life forms would be destroyed, rather than nurtured. We will have to wait to see how this controversy resolves itself, if it ever does.

MARS AND ANTARCTICA

In the few years since we began finding martian meteorites in Antarctica, we have cleared a number of conceptual hurdles. We have found martian atmospheric gases in impact-melted glass in EETA 79001 that proves the SNC meteorites came from Mars. We have come to the realization that gigantic impacts can transfer samples from some planets to other planets. We have described a variety of rocks that formed as a result of igneous activity on Mars and have been able to speculate about the parent magmas that produced them. We have found ourselves willing to consider the possibility that life began independently on another planet in the solar system. Perhaps we even perceive that the day is a little closer when we can anticipate meeting other intelligent beings, but, please, not *too* intelligent!

SUGGESTED READING & DISCLAIMER

McSween, H. Y., Jr. (1994) What we have learned about Mars from antarctic meteorites. *Meteoritics* **29,** 757–779.

McSween, H. Y., Jr. and Treiman, A. H. (1998) Martian meteorites. In *Planetary Materials*, vol. 36, *Reviews in Mineralogy*, ed., J.J. Papike. Washington, D.C.: Mineralogical Society of America.

Treiman, A. H. (1995) S ≠ NC: Multiple source areas for Martian meteorites. *Journal of Geophysical Research* **100,** 5329–5340.

Workshop on the Issue Martian Meteorites: Where do We Stand and Where Are We Going? Abstracts of 51 papers given at a workshop held Nov. 2–4, 1998. Compiled by Lunar and Planetary Institute personnel as Contribution

---

[4] This could be taken as proof of life on earth, because intelligent life would have had to be present to build the automobiles.

No. 956, Lunar and Planetary Institute. Available from Order Department, Lunar and Planetary Institute, 3600 Bay Area Blvd., Houston, TX 77058–1113, USA.

5 http://www.space.com/scienceastronomy/solarsystem/mars_split_ 010327.html This is an interesting website containing an extensive discussion of the pros and cons of the evidence for martian life.

5 The publisher has used its best endeavours to ensure that the URLs for external websites referred to in this book are correct and active at the time of going to press. However, the publisher has no responsibility for the website and can make no guarantee that a site will remain live or that the content is or will remain appropriate.

7   Meteorites from the moon

INTRODUCTION

It may be impossible to overestimate the importance of the moon to us earth-dwellers. The moon likely provided our first correct intimation of the idea that one body can orbit another. Eratosthenes showed us how to measure the diameter of the earth in the second century (BC), and made a pretty good estimate of it himself. With a refined measurement of the earth's diameter, we could use that number as a baseline to measure the distance from the earth to the moon. Knowing this distance, we did not find the numbers so incredible when we then measured the distances to nearby asteroids; then to Apollo's Chariot and to all the planets of the solar system, using essentially the same method with simple geometry. Knowing the distance to the sun allowed us to calculate distances to nearby stars, using the diameter of the earth's orbit as a measuring stick. It all started with measuring the distance to the moon.

But much more recently, the moon served as a stepping-stone of another kind. It was a nearby body. We might aspire one day to stand on the moon, because it was so close. We did so aspire, and brought it about. With the astronauts standing there, we thought many previously unthinkable things: among them, the long jump to Mars suddenly seemed within our capabilities. We will certainly accomplish that visit also, and this will lead us ever farther. Who can say how far?

Imagine if we had lived on an earth that had never had a moon – we would not have had a stepping-stone. A trip to Mars, while theoretically possible, probably would still be considered an outlandish idea. How would this have affected our philosophy and our science? There would be no comparative planetology today, that much is certain. How would this have affected our society? For better or for worse, we would today be much more inward-looking. The jump to Mars might occur much later in human history, or not at all. In a sense, the moon has turned us outward, into the Great Emptiness. It has also sent us some pieces of itself, which have fallen as meteorites and which we can find in Antarctica and other places, if we search patiently enough. We recognized our first lunar meteorite during the 1981–82 field season.

A LITTLE HISTORY

On January 14, 1982, we were camped at the Allan Hills Main Icefield. We had experienced a reasonable number of good-weather days, interspersed with periods of high winds and blowing snow, and a few days with no wind, but semi-whiteout conditions with thin ice fogs. John Schutt's field notes indicate that from December 25 through January 14 we had collected 282 meteorite specimens; 232 of them were found on the Main Icefield and 50 had been collected during several day trips out to the Allan Hills Near Western Icefield, about 25 km to the west.

On the evening of January 14, Ursula Marvin was throwing the canvas shroud over her snowmobile when the wind suddenly caught it. Ursula lunged for it, but got her foot tangled in the support struts for the front ski and went down with a broken leg. At the time, she didn't realize the bone was fractured and spent a painful night as the leg throbbed increasingly. By morning she knew her field season was over. She was fortunate in one sense, because we were expecting the arrival of Ian Whillans by helicopter from McMurdo. Ian, who died in 2000 at far too young an age, was a glaciologist from The Ohio State University. The helo arrived around 10 a.m. By that time, Ursula could not even hop on one foot, so John Schutt picked her up and carried her over to the helo. The Navy doctors in McMurdo put her leg in a walking cast for her return to Christchurch, New Zealand. Ian moved in where Ursula had been, and prepared to spend a few days.

We now had a glaciologist in camp. I had begun to think that meteorite stranding surfaces, aside from being unique sites at which to collect meteorites, might have special characteristics that could tell us things about the ice sheet, and I looked on Ian's presence as a good opportunity to consult a glaciologist on the spot. My field notes for January 15 indicate that three of our field team collected meteorites while the remaining two escorted Ian around the general area. In addition to soliciting his impressions about the local meteorite stranding surfaces, I wanted to get his opinion about the "cobble dunes" of Manhaul Bay.

Allan Hills has a vaguely Y-shaped appearance (see Frontispiece). The arms of the Y enclose a patch of bare ice that does *not* support a concentration of meteorites; we had found only one, and this on an area of ice almost half the size of the Allan Hills Main Icefield, which had already yielded some hundreds of specimens. Ice flowing northeastward off the ice plateau of East Antarctica detours past the Allan Hills structure, essentially making a left turn to flow north past the upstream arm of the "Y". Once past the barrier, it turns northeastward again, to be carried by the Mawson Glacier to the sea. In doing so, it passes the open arms of the Allan Hills "Y" with little or no detour into that embayment, whose name is Manhaul Bay. Manhaul Bay is filled with ice that is ablating faster than it is accumulating, particularly near the junction of the arms of the "Y." It is pleasant now and then to be able to get off the ice and wander around on bare rock. Many of us are geologists, after all. There is a lot of petrified wood scattered around on the sandstone surface; most of it highly polished by sandblasting, attesting to a certain degree of windiness in that particular zone. More to the point regarding wind, however, are lenticular accumulations of cobbles up to the size of a tennis ball, lying on the bare rock in piles about 20 cm high and 1–2 m long. These mounds have been described by Hal Borns, of the University of Maine. His best guess as to their origin is that they are wind-blown dune deposits. Imagine winds that pile cobbles into dunes! I showed Ian the cobble dunes and explained Hal Borns' ideas to him, hoping for a counter suggestion. He found them very interesting, striding around from one to the other, and I felt confident that he would very shortly produce an ingenious hypothesis to explain them. Finally he stopped by one of them and studied it for a very long time. I was staring at it too, hoping to divine what it was that set this one off from its fellows, and what subtle characteristic was telling Ian such volumes. Finally, I said,

Well, what do you think?

and Ian said,

I don't know, but I think we should get out of here before the wind comes up!

ANSMET FINDS A LUNAR SAMPLE

On January 16, we did some ice coring under Ian's direction and toured the Main Icefield some more. On the 17th we awoke to strong winds that abated slowly, so that we could work by afternoon. We carried out additional core sampling and mapped meteorite locations. By the 18th, Ian wanted to see the Near Western and Middle Western Icefields, so he went off with John while the rest of us collected meteorites at the Near Western site. He and John collected 11 specimens at the Middle Western Icefield. Ian found the first one. John's field notes describe it as follows:

#1422[1] – *Strange meteorite. Thin, tan-green fusion crust, ~50%, with possible ablation features. Interior is dark grey with numerous white to grey breccia (?) fragments. Somewhat equidimensional at ~3 cm.*

So the specimen was still about 50% covered by fusion crust and was regular in shape, with a diameter of roughly 3 cm (Figure 7.1).

There the matter stood until the sample reached Houston. It would be nice to report that we all had a look at it in the field and tentatively identified it as a lunar sample, but that did not happen. Reading

---

[1] #1422 is our field number. When a specimen is unpacked at Johnson Space Center a new number is assigned. The new number becomes part of the meteorite's name. This meteorite's name is ALHA 81005. It is a lunar sample.

FIGURE 7.1 Lunar meteorite ALHA 81005, as found on the ice at the Allan Hills Midwestern Icefield. Notice the cm scale on the black box, which also records the specimen's field number. (Photo by Ian Whillans.)

John's description, though, one wonders why. I have to conclude that none of us had an imagination that was stretchable that far.

It was a different story at the Johnson Space Center when the sample was unwrapped and christened ALHA 81005 (Figure 7.2). These people had been working with lunar samples for years and their first reaction on looking at it was that this was a lunar sample. This feeling quickly developed into deep conviction. They were as good at recognizing rocks from the moon as we had become at recognizing meteorites on the ice. We had recognized a meteorite and they had recognized a lunar sample! Now the Meteorite Working Group (MWG) would have to deal with it.

The MWG is an advisory group whose members are scientists doing research on meteorites. Since the second year of the ANSMET project, the MWG has met two or three times per year to assess all the requests that have been received for research material from the ANSMET collection. Their purpose is to make recommendations

1 cm

FIGURE 7.2 Lunar meteorite ALHA 81005 in the Antarctic Meteorite Laboratory at NASA's Johnson Space Center. The sample has been partially chipped away to reveal a fresh surface. Notice the angular shapes of the white inclusions. These anorthosite grains are fragments of broken rock that were embedded in fine-grained debris on the lunar surface. This aggregation was then compacted to form rock during impacts on the moon. The texture is described as a *breccia*. (NASA photo.)

about the most effective ways to get antarctic meteorite samples to the scientific community, while preserving as much as possible of the original specimen for future use. In a typically bureaucratic arrangement, their recommendations are made to a Meteorite Steering Group (MSG), consisting of three members: one from NASA, one from the Smithsonian and one from the National Science Foundation. The MSG then decides whether or not to grant the recommended material, and directs the Antarctic Meteorite Laboratory at the Johnson Space Center to carry out a sampling program to supply the material.

The MWG performs a valuable function, because antarctic meteorites are typically small and are often unique, or rare enough to be deemed very precious. Even-handed consideration of research requests often can help to conserve material. ALHA 81005, for example, was about the size of a golf ball and weighed only 31.4 g (Figure 7.2). If it were a lunar sample, it would be the first indication that impacts on the moon actually could deliver lunar rocks to the earth, and it would support the concept that the SNC meteorites could have been derived from Mars by a similar mechanism. On the other hand, if it were shown to be non-lunar it would be a completely new, previously unknown type of meteorite. Either way, ALHA 81005 was bound to be interesting and it generated much intellectual salivating in the community.

THE MWG GETS INTO TROUBLE

Rumors brought news of the find before the official announcement appeared, and some members of the community heard about it before others. Since the specimen was so small, a logical approach that would conserve material could be to distribute samples to consortia – groups of scientists with a variety of research specialties – organized to carry out *sequences* of studies using the same small sample. Through the rumor mill, part of the community got a start ahead of their colleagues in informally arranging consortia, in order to be ready to request material when the news became official. Scientists who heard about it somewhat later often found that colleagues had already been snapped up by competing consortia, and at this point the situation became a lot more complicated.

There was a sudden feeling of distrust in the air and a vague perception of shadowy hierarchies of people who were "in" having an advantage over some who had thought they were "in" but now suddenly getting the feeling they were "out." This can be unsettling, and the MSG received an outspoken letter from a respected member of the planetary community, complaining bitterly about the situation. He had heard the news partway along the rumor chain. This seemed to be a case of someone who suddenly felt only "halfway in" (an optimist), or "halfway out" (a pessimist). On the other hand, the letter did contain a valid discussion of problems with the meteorite allocation system that until now had seemed to serve the scientific community quite well.

I was chairman of the MWG at that time, and had to admit that indeed we had not taken extra measures to release the news of this important new find everywhere simultaneously. Live and learn. What to do now? What had happened could not be reversed, but in order to minimize continuing criticism, I decided not to consider research requests for ALHA 81005 within the regular MWG.

This was in part a logistical decision. Many of the MWG members could be expected to request part of the sample for their own research and the committee's decisions would be weakened by the fact that key members would be absent during our consideration of their requests, as well as all competing requests. Instead, I would form an *ad hoc* committee composed of scientists who agreed ahead of time not to request material from this meteorite. With such a committee there would be no members absenting themselves during key discussions and no question of a conflict of interest in our recommendations. I found nine other highly qualified people who agreed to these conditions and asked Donald Bogard, of the Johnson Space Center, to serve as non-voting chairman of the committee. I have always had great admiration for Don as a scientist and also as one who can be completely impartial in his judgments. The *ad hoc* committee first met on December 4, 1982.

With the very strong suspicion that this meteorite was a lunar sample, the first order of business for the *ad hoc* committee was to favor research requests that tried to settle the question of its origin. The committee had a number of requests for sample material, and most of them were in the form of consortia. It was decided to postpone consideration of consortium efforts until we had a better idea of whether or not the sample was lunar. As a result, even though overall consortium requests were not to be considered, consortium members became eligible to receive samples as individuals if their part in their consortium seemed likely to help in answering the origin question. Certainly, the mineralogy of the sample and the fabric of the rock could be compared directly with those characteristics in lunar samples, so four thin sections were made and passed around to seven investigators for petrographic descriptions.

Some years before, Bob Clayton at the University of Chicago had discovered that meteorites had variable oxygen isotopic ratios. His measurements of many meteorites, terrestrial rocks and lunar samples showed that all these rocks could be separated into groups, with each group having a range of oxygen isotopic ratios not shared by other groups. Such measurements can serve to narrow the range of objects with which a particular rock can share a common origin. It would be crucial to know if the oxygen isotopes of ALHA 81005 lay within the earth–moon oxygen-isotope group, so 50 mg of sample material was allocated for that study.

Other studies that had been proposed, and for which the committee allocated material, were noble gas measurements, cosmogenic nuclides, nuclear particle tracks in feldspar grains, neutron activation analysis and rare-earth element analyses. All of these studies were thought to have the potential of differentiating lunar samples from asteroidal meteorites. In addition, the committee permitted a non-destructive reflectance study of a polished surface that might be able to differentiate between different source areas on the moon if, in fact, the sample turned out to be lunar. We permitted this because the reflectance spectrometer was located at Johnson Space Center and the measurement could be carried out during preparation of part of the sample for cutting thin sections. The committee also solicited two studies that had not been proposed: measurement of the magnetic properties of several small chips, and passive counting of a major piece of the specimen for aluminum-26(^{26}Al) measurement. ^{26}Al, an unstable isotope of aluminum produced by radiation in space, is measured routinely on most of the meteorites returned from Antarctica. It can help to determine the length of time during which a meteorite has existed as a small body in space or how long it has been on the earth's surface.

The list of investigations described above may suggest a profligate use of available material. Actually, the largest allocation for a single procedure was 480 mg, and the total of all allocations was around 2.6 g of this 31.4 g meteorite. About 4 g was also lost as tiny crumbs and cutting debris. To me, this is an amazing accomplishment. I can remember when the only research one would do on a rock sample was to identify the minerals optically in a thin section, make a modal analysis by point-counting a thin section, and get a bulk chemical analysis by sacrificing 10 g or so to destruction by wet chemistry. This would produce an analysis of perhaps 10 major elements. Nondestructive electron microprobe analysis of the individual minerals in a polished thin section of rock was unknown. Neutron activation analysis to measure concentrations of trace elements in the parts-per-billion, and sometimes parts-per-trillion range was unheard-of. Mass spectrometric analysis of samples to determine isotope ratios was only a dream. I compare *then*, with *now*, and cherish the feeling of wonder at what modern analytical techniques can achieve, with very little loss of sample material. These techniques have become crucial to the study of meteorites. When supplemented by the careful handling and documentation procedures used in the Antarctic Meteorite Laboratory at the Johnson Space Center, and the combined wisdom of the eight or nine specialists in various fields who donate their time to the MWG, these techniques reach their highest level of effectiveness.

RESULTS OF THE FIRST GO-ROUND

Every year in March, NASA sponsors a meeting called the Lunar and Planetary Science Conference. By March, 1983, most of the samples allocated at the December 4th MWG meeting had yielded results, the investigators had talked to one another, and it was generally conceded that ALHA 81005 is a lunar sample. More specifically, it is a polymict anorthositic breccia, i.e., it is a rock formed from broken fragments of other rocks, whose predominant mineral was anorthitic – i.e., rich in the $CaAl_2Si_2O_8$(anorthite) molecule. Most of the rock and mineral fragments looked like lunar highlands material but there were rare inclusions of rock fragments that looked like lunar mare basalt. These investigations benefited greatly from the fact that we already had "ground truth" information about the moon's surface in the form of samples collected by the astronauts. A series of formal presentations at the Lunar and Planetary Science Conference made the lunar connection quite clear. Briefly, the evidence was as follows:

- Fabric data.
  Depending on the sample, 30–55% of the rock is made up of clasts (rock fragments) some of which have been highly shocked but some of which show little or no shock effects. The fine-grained matrix contains mineral fragments, irregular-shaped brown to colorless inhomogeneous glass fragments, glass spherules and shock-melted soil masses. Typically for lunar breccias, the clast fragments are bits and pieces of a variety of rock types known to be present on the moon. *In every detail, the rock looks like a lunar breccia.*

- Mineralogical data.
  Highly calcic plagioclase, anorthite, is the major mineral in the clasts, the matrix and the glass compositions. It is present in a modal abundance of ~75%. No known meteorite is so highly anorthositic, but this is typical of lunar highland regolith breccias. The second most abundant mineral, pyroxene, has a range of MnO/FeO (manganese oxide/ iron oxide) values that is consistently lower than those ratios in basaltic achondrites, a type of meteorite that superficially might seem to be closest to ALHA 81005. *The MnO/FeO ranges of the pyroxenes are completely consistent with those values for lunar breccias.*

- Compositional data.
  Bulk chemical analyses show that ALHA 81005 is inconsistent in composition with both terrestrial rocks and other known meteorite types, but is characteristic of lunar material. These characteristics include the relative abundances of the rare earth elements and certain concentration ratios such as low

MnO/FeO, high Cr/Fe (chromium/iron), high Al/Fe (aluminum/iron), and low Na/Ca (sodium/calcium), K/Ca (potassium/calcium), Rb/Ca (rubidium/calcium) and Cs/Ca (cesium/calcium). *It is similar to lunar samples in having a very low content of volatiles and very low concentrations of a group of elements called chalcophiles that includes sulfur and elements that are geochemically similar to sulfur.*

- Oxygen isotope data.

ALHA 81005 has an oxygen isotopic composition that lies on the earth–moon fractionation line and is typical of regolith and fragmental breccias inferred to be from the lunar highlands. *The oxygen isotopic composition distinguishes this sample from eucrites, which are the other high-Ca meteorites.*

- Noble gas data.

ALHA 81005 contains solar wind-implanted gases. *The absolute and relative concentrations of these gases are quite similar to lunar regolith samples but not to other meteorites.*

- Cosmic ray exposure history.

ALHA 81005 contains the cosmogenic nuclides ^{10}Be (beryllium-10) and ^{26}Al at levels that suggest the time spent in space by this meteorite is shorter than that spent by most meteorites whose orbits have aphelia in the asteroid belt. *This short exposure time is consistent with a lunar origin for this meteorite.*

- Magnetic properties.

The ferromagnetic resonance spectra of ALHA 81005 are similar to those of the lunar soil, and the total concentration of metallic iron in the sampled chips is within the observed range for Apollo 16 rocks and soils. *Magnetically, the specimen looks like a lunar sample.*

- Nuclear particle tracks.

Bombardment in space by energetic nuclear particles (heavier than calcium) will break chemical bonds, producing an effect similar to very local melting as they plunge through the crystalline minerals of a meteorite. Each such particle leaves a track, in the form of a microscopic glass-like channel. The presence of these channels can be demonstrated by mild etching with acid, because the glass is more soluble than the crystalline structure surrounding it. Materials collected from the actual surface of the moon, as well as meteorites smaller than about 3 m in diameter, show nuclear particle tracks because of their exposure to high-energy cosmic rays. *Clasts in ALHA 81005 had no detectable nuclear tracks, and therefore could not have been exposed at the exact surface of its parent body or have had nearly as long a sojourn in space as an asteroidal meteorite.*

- Thermoluminescence data.

When radioactive elements decay or low-energy cosmic-ray irradiation occurs, defect sites are produced between atoms in a crystalline structure. Stray electrons can take up residence in these sites and can be released again by heating. With such release, faint flashes of light occur. These flashes are called thermoluminescence (TL) effects. Over a long period of time, natural radioactivity will cause a buildup of defect sites in minerals; particularly in feldspars, of which anorthite is one. The concentration of defects tends toward an equilibrium level, because the sites will also "heal" spontaneously at a slow rate. If the mineral is heated to ~300 °C all the accumulated electrons will be discharged over a small temperature interval as the defects are healed. As each defect readjusts to a lower energy level it releases energy in the form of a tiny flash of light, and the accumulated intensity of the light flashes will be a measure of the total number of defects that had been present.

The TL of ALHA 81005 was very weak. Since the minerals could be assumed to be ancient, and therefore should have produced a strong TL signal, it seemed likely that any early TL had been bleached out by shock heating – presumably the shock that ejected the specimen from its parent. The low TL level observed must have been induced during its sojourn in space as a small body. The weak TL response constrained this period to less than

2500 years. While some asteroidal meteorites fall with similarly low TL levels, the great majority of them have much higher levels of TL, consistent with cosmic ray exposure ages of tens of millions of years. *So the TL data of ALHA 81005 are more consistent with origin from a nearby body such as the moon, rather than with a meteorite parent body.*

Statements culled from published reports by the original investigators of ALHA 81005 are quoted below. They reflect their depth of conviction that this specimen is a sample of the moon.

MnO/FeO ratios in pyroxene, texture (abundant brown, swirly glass, in general typical of lunar regolith breccias), and overall composition ($\sim$ 75% plagioclase) all indicate that Allan HillsA 81005 is of lunar origin.

Petrographic data and pyroxene compositions indicate that meteorite ALHA 81005 is a breccia from the terrae (highlands) of the Earth's Moon.

Petrography and mineral compositions show that ALHA 81005 is a lunar highland regolith breccia and is the first known lunar meteorite.

Meteorite ALHA 81005, a glassy regolith breccia, is beyond any reasonable doubt of lunar origin, according to its petrographic, chemical, and isotopic characteristics.

Meteorite ALHA 81005 is a shock-compacted lunar highland regolith breccia.

The antarctic meteorite ALHA 81005 has oxygen and silicon isotope ratios identical with lunar highland rocks.

The noble gas data alone are strong evidence for a lunar origin of this meteorite.

Antarctic meteorite ALHA 81005 is a regolith breccia apparently sent to earth by an impact event in the lunar highlands.

ALHA 81005 is the first meteorite to be decisively identified with a specific parent body.

These (element) ratios and the general similarity of the chemical composition of ALHA 81005 with lunar highland rocks do not leave any doubt as to the lunar origin of this antarctic meteorite.

...ALHA 81005 exhibits those features evident in lunar highlands samples returned by the Apollo missions and we conclude that it is, in fact, of lunar origin.

Based on the well-established characteristic lunar and meteoritic ratios of FeO/MnO, Cr_2O_3/V and K/La, and REE (rare earth element) patterns, ALHA 81005 meteorite is undoubtedly of lunar highland origin.

In all respects ALHA 81005 is more similar in composition to materials returned by the Apollo and Luna missions than to other meteorites, typical terrestrial materials, or all non-lunar materials of which we are aware.

The Fe/Mn ratio, along with petrographic data, indicate that ALHA 81005 originated in the lunar highlands.

Based on incompatible element abundances, as well as other studies in this volume, ALHA 81005 is clearly a lunar sample.

At the end of a long session devoted to ALHA 81005, the last paper was given by Randy Korotev, a well-known member of the Washington University lunar study group. By that time, the results had gotten somewhat redundant, so he started his comments by saying, "I'd like to present some evidence for why we believe ALHA 81005 is *not* from the moon." A sudden hush enveloped the audience and Randy had everyone's undivided attention. Then he continued,

"Unfortunately, we were unable to find any such evidence so I'll have to talk about something else..." The audience roared.

The remarkable result of the first round of investigations of this meteorite is that among a community of scientists known for its skepticism and its often conflicting interpretations of scientific evidence, there was no debate. Without the slightest shadow of doubt, this had to be a lunar sample.

With identification of ALHA 81005 as a lunar sample, a milestone had been passed. We now knew that impacts on the moon actually could deliver lunar rocks to the earth, and, more importantly, we now had support for the concept that the SNC meteorites could have been derived from Mars by a similar mechanism. To quote Ursula Marvin (she of the broken leg), this 31.4 g rock,

> ... had profoundly altered our views of the types of planetary materials we may expect to find here in the future. Its mild degree of shock is one of the most surprising aspects of this specimen. High energy meteorite impact on the moon is generally accepted as the only process capable of accelerating target materials above the lunar escape velocity of 2.4 km/sec, and, up until now, the only materials expected to go into orbit were those excavated from depth below ground zero and shock-melted to glass. Contrary to all predictions, Allan Hills 81005 came from the shallowest layer on the moon, the surface soil, and it shows no greater shock effects than those in many samples the astronauts lifted off the lunar surface with tongs. This specimen has generated a new round of calculations on the dynamic effects of accelerating lunar ejecta into earth-crossing orbits.

Furthermore,

> If unmelted materials can be ejected from the surface of the moon, can they be also from the surface of Mars?

Well put, Ursula!

At this point, ALHA 81005 had given us important insights into impact dynamics and the origin of the SNC meteorites. It was time for it to tell us something about the moon. We were now ready for a second round of sample allocations. The papers quoted above had all been presented on March 18, 1983, and that evening the *ad hoc* committee met again to consider all earlier requests that had been postponed, as well as all requests that had been received since the December 4, 1982 meeting. Now the situation was different. As two investigators, Allen Treiman and Michael Drake stated, "... the meteorite has, in effect, provided us with another mission to the moon ...," and now we knew we were distributing lunar samples.

The procedure in studying a lunar polymict breccia is to treat it as many samples of different rock, each of which has its own story to tell about its parent magma and how long ago it crystallized. Therefore the allocation plan now adopted by the *ad hoc* committee emphasized the distribution of individual rock fragments, separated from the breccia matrix. The committee's recommendations took final form over the course of a single evening. In assessing the research proposals, we had to consider the following, already-existing, knowledge about the tortured face of the moon, and the moon's history.

WHAT WE ALREADY KNEW FROM REMOTE SENSING STUDIES AND APOLLO SAMPLES

Space weathering and lithification

Even before we first visited the moon, there was evidence that it gets darker over time. Large impacts excavate fresh material from deep below the surface and throw it out in ray-like patterns around impact craters. This crushed ray material is bright around the freshest-looking craters, darker around the older-looking craters and invisible around the oldest craters, apparently having merged into the general drab grey color of the lunar surface.

The astronauts had collected bags of the lunar "soil." Lunar soil or, more properly, lunar regolith, consists of the fine comminution products of the near-surface rocks and bits of glass from large impacts, as well as from a myriad of micrometeorite impacts. Lunar soil is subject to several agents of change that, in combination, cause it to darken and to lithify (i.e., become rock-like).

Energetic particles making up the solar wind, mainly electrons and hydrogen and helium nuclei streaming outward from the sun at about 500 km/s, bombard the surface. Solar wind bombardment causes an effect known as sputtering; this is the knocking loose of individual atoms from exposed surfaces. Some fraction of these atoms can escape from the moon but most strike other surfaces first, and are deposited there as "sputter coatings." The most abundant element in lunar rocks is oxygen, and a larger fraction of sputtered oxygen escapes the moon than of heavier elements. The net effect of this preferential escape of oxygen is preferentially to concentrate the heavier elements in sputter coatings. Chief of these, among the more common elements, is iron, with 3.5 times the mass of the oxygen atom, so sputter coatings on surface grains are enriched in iron, which aggregates into grains several nanometers in size. Sputter coatings help bind the fine grains in lunar soil together, in a lithification (rock-forming) process.

Another process that helps in the rock-forming sequence is micrometeorite impact, which not only creates glass-lined microcraters on rock surfaces but also causes melting and thermal vaporization of soil target materials. The vapor tends to be injected downward, so micrometeorite impact vapors are forced into the upper cm or two of the lunar soil, where they condense on grain surfaces and assist in welding them together. Again, however, oxygen present in that fraction of the vapor that is ejected upward can escape more readily from the moon than can heavier elements.

Thus, over a period of time, submicrosopic grains of metallic iron accumulate in the upper few millimeters of lunar soil due to fractionation by sputter deposition and dissociation of iron oxide in the vapor phase produced by micrometeorite impacts.

These are the ways in which rock-forming processes act on the lunar soil, and the result is cindery-looking objects sometimes described as soil breccias, and, if the soil contained embedded, broken rock fragments, as fragmental breccias. The general term describing these newly formed rocks is agglutinates. Agglutinates tend to be mixed downward into the regolith by the gardening effect of macrometeorite impacts. Agglutinates are rather dark in color, as is the normal surface soil of the moon. This darkening was originally thought to be caused by the large fraction of glass present, which appeared darker than the minerals from which it formed. The fundamental cause of the darkening effect, however, is optical absorption by submicroscopic iron particles liberated by the sputtering and vapor deposition described above. As these processes of space weathering proceed, fresh, highly reflective, light-colored material thrown out as ejecta around recent craters becomes slowly darker.

Mixing of rock types, and inferences from this

Due to its bombardment history, recorded by the density of impact craters of all sizes, the surface of the earth's moon at any one point is a hodge-podge of materials from many sources. A crater-forming impact in one place will scatter target rocks of one type to many other points surrounding it, to lie next to possibly different rock types derived from other impact craters at other places. Rocks can be tossed hundreds of kilometers across the surface, so finding a lunar rock on the surface does not mean the rocks below are of that type.

Shock compression and melting from later impacts can lithify masses of fragments, to produce new rocks called impact-melt breccias and regolith breccias that consist of assorted fragments from different places on the moon.[2] One would expect, however, that rock fragments found in a lunar impact breccia would consist *mostly* of locally derived rocks, with lesser numbers of fragments representing more-distant impacts. From this we could expect to infer the general underlying rock type in any area in which we could pick up samples.

---

[2] By definition, a breccia is a rock formed by lithification of broken bits of rock. A rock that has been crushed and relithified is termed a *monomict breccia*. A rock formed by lithification of fragments from different sources is termed a *polymict breccia*.

The American and Russian missions had sampled rocks disproportionately in mare areas; the main rock type found in those breccias had been generally basaltic in composition, so we inferred that
the maria were composed of basaltic rocks. Basaltic rocks are quite
common on the earth, originating as lava flows and as intrusive rocks
whose lava did not reach the surface before cooling and crystallizing.
The small crystal sizes in the lunar basaltic rocks indicate that they
cooled rather quickly. This suggests that they originated as lava flows
on the surface.

Less frequent among the lunar impact breccias from the Apollo
program were fragments of a rock type rich in a calcium–aluminum–
silicate mineral called anorthite, but poor in the magnesium-rich silicates that we would expect to crystallize from a magma at the same
high temperature as anorthite. Coarsely crystalline anorthite would
be expected to have crystallized at some depth below the surface.
At depth, temperatures would be much higher and cooling would be
slower. Crystals would grow larger, and there would be an opportunity for the co-crystallizing magnesium-rich silicates to separate from
the anorthites by density stratification. This is often the case, for example, for terrestrial igneous rocks of deep-seated origin which have
been exposed at the surface only after erosion has stripped away the
overlying materials.

The Apollo 11 astronauts landed in Mare Tranquillitatis (Sea
of Tranquillity), some distance away from the nearest lunar highlands area. The samples they collected contained minor fractions of
anorthite-rich rock fragments which, from their small sizes and sparse
occurrence in the bulk collection, were thought to have been projected
from sites distant from the landing site. "Distant" could mean the lunar highlands and, in fact, many workers speculated that the highlands
were indeed the source for these lighter-colored rocks.

Although anorthositic rock fragments have been found at every
site we have visited on the moon, those rock fragments that are rich-
est in the anorthite component, the so-called "ferroan anorthosites,"
were common only at the Apollo 16 site, which, of all our lunar

landing sites, we believed to have the highest contribution from the lunar highlands. This supported the idea that anorthite-rich rocks would be found in great abundance in the lunar highlands and, curiously, that the lunar highlands have surface rocks of a type that would be expected to form only at great depth.

Agreeing with this conclusion, a group at the Smithsonian Astrophysical Observatory including John Wood, John Dickey, Ursula Marvin and Benjamin Powell went one step further and proposed that the lunar highlands, which occupy a much larger fraction of the moon's surface than the maria, consist completely or predominantly of anorthite-rich rocks. They proposed that a moonwide igneous fractionation process had caused lower-density anorthite crystals to float upward to form the lunar highlands, and higher-density magnesium-rich silicates to sink, and concentrate downward. Because this must have happened on a moonwide scale, in order to produce the massive accumulations of anorthite-rich rocks making up the lunar highlands, the unique requirement of this model was that intense heating and global melting of a significant fraction of the moon must have occurred early in lunar history. This was an unsettling concept because no one was sure what could have caused heating to such a degree, and we didn't think it had occurred early in the history of its sister planet, Earth. This model came to be referred to as the "moonwide magma ocean." Discussion of its merits and implications saw some heady days, indeed.

While the most common surface rocks on the moon are polymict impact breccias, occasionally a "pristine" rock is encountered in the Apollo collections. A pristine rock is an igneous rock that has not been mixed mechanically with other rocks and soil. It is thus a true representative of the original igneous crust, uncontaminated by addition of materials from other sources. Pristine lunar rocks may have been brecciated, but they still consist of only one rock type; therefore they can give us a clearer picture of a single stage in the igneous history of the moon. Polymict impact breccias, because they may contain bits and pieces of many rocks from many different sites, also may contain inclusions that themselves are pristine rocks. Most of the pristine rocks and inclusions we have collected on the lunar surface represent three general rock types. These are: (1) igneous basalts from the lava flows that filled the mare basins; (2) anorthite-rich igneous rocks which we believe accumulated as a scum at the surface of the moonwide magma ocean and which now form most or all of the lunar highlands; and (3) a series of rocks richer in magnesium and iron that we can understand as resulting from the kind of igneous rock-forming processes that occur on the earth within magma chambers of relatively limited size. These last are called *magnesium-suite* rocks (Figure 7.3), and we assume they come also from the lunar highlands.

Implicit in the concept of a primordial moonwide magma ocean is the notion that initially the moon was completely covered in anorthite-rich rocks identical to the lunar highlands of today. Early in the moon's history, however, it suffered a terrible bombardment by large bodies that blew tremendous holes, i.e., the mare basins, into the already formed anorthite-rich crust. This actually may have been going on all during the formation of the moon as it accumulated planetesimals, but if so, all we see are the results of the last great collisions, which erased the earlier ones. These collisions released so much energy by shock compression that some of the underlying deep mantle rocks, too deep even to be melted during the magma ocean phase, may have been partially melted. The rock that remained unexcavated below these great impact cavities was shattered, and provided easy conduits through which these partial melts could escape as floods of lava, running out onto the surface and filling the vast impact basins to form what came to be called the lunar maria (seas). The last extrusive volcanic activity whose age we have measured occurred about 3.1 billion years ago, but based on a low density of impact craters, some mare surfaces may be as young as 2 billion years. From the time of the great lava floods to the present, the moon has been eerily quiescent: it is geologically dead. Asteroids have crashed onto its surface, of course, and have sent bits and pieces of the

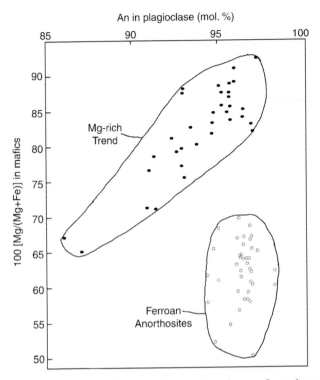

FIGURE 7.3 The term "An in plagioclase" reflects the variable composition of the calcium–sodium–aluminum silicate called plagioclase, in which *an*orthite is the calcium-rich end. It really reflects the following ratio: calcium/(calcium + sodium) in the plagioclase molecule. The important thing to notice is that, based on composition, the lunar highlands rocks and minerals appear to consist of two generic families: ferroan anorthosites and magnesian-rich rocks. The term "mafics" refers to rocks rich in magnesium (Mg) and iron (Fe). Reproduced with permission from Warren, P.H. and J.T. Wasson (1979) *Proc. Lunar Planet. Sci. Conf. 10th* p. 604, *Geochimica et Cosmochimica Acta* (Supplement 11), New York: Pergamon Press.

moon spinning into space, but the craters they formed have been relatively puny compared to the mare-forming impact basins, and the response to these lesser impacts has not been flooding by extruded lava. The only other recent disturbances have been a few space vehicles intentionally crashed onto the surface and the pitter-pat of astronaut feet.

NEW DIMENSIONS FROM THE LUNAR METEORITES

The Apollo astronauts, plus the Russian Luna vehicles, were sent by nervous engineers to locations that appeared to present benign landing sites, free of surface irregularities that could endanger the success of the mission. These tended to be sites in the lunar maria, and so most of the Apollo samples are predominantly mare samples. Also, most of the sites were close to the equator and on the lunar nearside. As a result of this, all of our available lunar samples had been collected within a sampling area of only 4–5% of the lunar surface and are more representative of the maria than of the highlands. Asteroids, on the other hand, don't care where they strike the target, so our lunar meteorite, plus the others that soon would be found, are giving us a much wider sampling of the moon, including probably the lunar farside. *This is good.*

Remember that ALHA 81005 is a polymict breccia. It contains rock fragments from many different sources, so essentially it is many rock samples in one small package, bound together by lithified lunar soil. Study of this single sample therefore yields information on a number of different lunar rock specimens. *This is good.*

Because ALHA 81005 contains many more anorthosite fragments than mare basalt fragments, it is believed to have come from a lunar highlands terrain, probably from a site rather distant from any mare source. However, we don't know exactly where. *This is unfortunate.*

Since the two "goods" seem to outweigh the single "unfortunate," and the "goods" are in reality "*very* goods," the availability of a lunar meteorite from Antarctica, on balance, was exciting news.

LATER DEVELOPMENTS

From 1981, with the discovery of ALHA 81005, through to 2002, seven additional lunar samples were recovered by the ANSMET program in Antarctica. All of these have been allocated by the MWG in a much less tension-filled atmosphere, with no need for a special *ad hoc* committee. Also, it was no longer necessary to undergo a rigorous process to prove a lunar origin, such as the one that had been required with ALHA 81005.

It had been a very lucky circumstance that the highly qualified technicians who first unwrapped ALHA 81005 at the Johnson Space Center were the same people who previously had unpacked and processed so many lunar samples, because *they* are the ones who immediately recognized the specimen as possibly lunar. Realizing now that meteorites could be lunar samples, the Japanese, who had a parallel program for meteorite collection in Antarctica, had a second look at some of their more puzzling finds and discovered *three lunar samples* in their collection that had been picked up during their 1979–80 expedition to the Yamato Mountains.

As of Spring, 2002, we are aware of the existence of 35 lunar meteorites (Table 7.1). Eighteen have been found in Antarctica: eight by the ANSMET program and 10 by JARE (the Japanese Antarctic Research Expeditions). In addition to the antarctic specimens, private collectors have found one lunar meteorite in a collection of meteorites from the central desert of Australia and fifteen in desert areas of North Africa and the Arabian Peninsula. Particularly fertile ground for lunar meteorites, apparently, is the Dhofar region of the Kingdom of Oman, on the Arabian Peninsula. About a third of the world's 35 lunar meteorites have been found there. Based on preliminary examination, however, Mahesh Anand, at the University of Tennessee, suggests there could be extensive fall-pairing of Dhofar specimens.

COMMENTS ON TABLE 7.1

- The table has two major divisions: an upper one, in which complete data exist and a lower one, containing lunar meteorites that have been recovered so recently that data are uncertain or nonexistent. Later discussions of fall-pairing and site-pairing rely on the existence of complete data; therefore the following discussions deal only with the upper part of the table.
- In the column labeled "Name," Y = Yamato Mts., A = Asuka, EET = Elephant Moraine, QUE = Queen Alexandra Range,

ALHA = Allan Hills, DaG = Dar al Ghani (Libya), MAC = MacAlpine Hills, NWA = Northwest Africa. Calcalong Creek is a location in Australia and Dhofar is a location in Oman. This column lists all the known lunar meteorite specimens, current through Spring, 2002. Only two specimens, Y 793169 and A 881757, are pristine rocks.

- In the column headed "Probable source ", all samples except Y 793169 and A 881757 are impact breccias of various kinds, classified as mare or highlands specimens based on the relative proportions of their mare and highlands components. Actually, QUE 94281 and Y 793274 are around 50/50 mare/highlands, so it is difficult to decide their probable provenance. Grouping them with the three other mare samples EET 87521, EET 96008 and NWA 032, is perhaps more of a guess than are some of the other grouping suggestions.

- In the next three columns, estimates are given of the travel times from the moon to the earth, time since fall on the earth and times of ejection from the moon, all in millions of years. The numbers for ejection from the moon can be seen to equal the sum of travel time and time since fall, incorporating the uncertainties in both values.

- The column labeled "Terrestrial pairing" indicates which lunar meteorites may have fallen as fragments of a single body. Members of a fall-pair are petrographically similar, they were found close together, they appear to have been ejected from the moon at the same time and appear to have been on earth for the same length of time. The implication must be that they are parts of a meteorite that was ejected from the moon as a single specimen that disrupted only during passage through the earth's atmosphere, or upon striking the earth's surface; therefore they can be thought of as a single meteorite in the conventional sense. Counting each pair or triplet as one specimen, plus the single meteorites that are not fall-paired gives the maximum number of sites (15) that the lunar meteorites for which we have good data can represent on the moon.

Table 7.1. *Lunar meteorites known as of Spring, 2002. Horizontal lines separate groups that may be site-paired. Site-pairing judgments are based on apparent equal ejection times from the moon, plus general petrological similarity. Because of insufficient data, site-pairing suggestions cannot be made for any specimen listed below Dhofar 025 in the Table. Contributors to the data in this Table include Otto Eugster, A.J.T Jull, Kunihiko Nishiizumi and Mahesh Anand.*

Name	Original Mass (g)	Probable Source	Moon to Earth (Myr)	Time on Earth (Myr)	Ejection from Moon (Myr)	Terrestrial Pairing	Lunar site Pairing
Y 793169	6.07	Mare	0.9–1.3	<0.05	0.9–1.3		Site-paired
A 881757	442.12	Mare	0.8–1.0	<0.05	0.8–1.0		Site-paired
EET 87521	30.7	Mare	0.002–0.004	0.05–0.11	0.05–0.11	Fall-paired	Site-paired
EET 96008	53	Mare	0.002–0.004	0.05–0.11	0.05–0.11	Fall-paired	Site-paired
QUE 94281	23.37	Mare	<0.1	0.02–0.08	0.02–0.18		Site-paired
NWA 032	300	Mare	0.039–0.045	0.0–0.080	0.04–0.12		Site-paired
Y 793274	8.66	Mare	<0.01	0.02–0.04	0.2–0.05		Site-paired
Y 791197	52.4	Highland	0.002	0.03–0.09	0.03–0.09		Site-paired
ALHA 81005	31.39	Highland	0.0025	0.04–0.10	0.04–0.10		Site-paired
QUE 93069	21.42	Highland	0.13–0.17	0.008–0.012	0.14–0.18	Fall-paired	Site-paired
QUE 94269	3.15	Highland	0.13–0.17	0.008–0.012	0.14–0.18	Fall-paired	Site-paired
DaG 262	513	Highland	0.07–0.09	0.07–0.09	0.14–0.18		Site-paired
Y 82192	36.67	Highland	5.1–11.1	0.048–0.118	5.1–11.2	Fall-paired	Site-paired
Y 82193	27.04	Highland	5.1–11.1	0.045–0.105	5.1–11.2	Fall-paired	Site-paired
Y 86032	648.43	Highland	8.0–12.0	0.042–0.102	8.0–12.1	Fall-paired	Site-paired

MAC 88104	61.2	Highland	0.01–0.08	0.21–0.25	0.22–0.33	Fall-paired	Site-paired
MAC 88105	662.5	Highland	0.01–0.08	0.21–0.25	0.22–0.33	Fall-paired	Site-paired
DaG 400	1425	Highland	0.20–0.24	0.016–0.019	0.22–0.26		Site-paired
Calcalong Creek	19	Highland	2.0–4.0	<0.03	2.0–4.0		
Dhofar 025	751	Highland	13–20	0.5–0.6	13.5–20.6		
Y 1153	?	Highland	?	?	?		
Y 981031	186	Mare?	?	?	?		
Y 983885	289.71	Highland	?	?	?		
Dhofar 081	174	Highland	~0.00?	0.20?	0.20?		
Dhofar 026	148	Highland	2–10	?	?		
Dhofar 280	251.2	Highland	?	?	?		
Dhofar 301	9	Highland	?	?	?		
Dhofar 302	3.83	Highland	?	?	?		
Dhofar 303	4.15	Highland	?	?	?		
Dhofar 304	10	Highland	?	?	?		
Dhofar 305	34.11	Highland	?	?	?		
Dhofar 306	12.86	Highland	?	?	?		
Dhofar 307	50	Highland	?	?	?		
Dhofar 489	34.4	Highland	?	?	?		
Dhofar 287	154	Mare?	?	?	?		

- The column labeled "Lunar site pairing" groups all lunar meteorites whose ejection times cannot be distinguished and whose probable source (mare or highlands) is the same. Members of the same site-pair group may have different earth-residence times and different transit times from the moon to the earth. Since they have identical or overlapping times of ejection from the moon, however, they could have been ejected from the moon during the same impact event; if so, they came from the same spot on the lunar surface even though they may have followed different trajectories in getting here. A secondary criterion is that all members of a site-pair be approximately the same kind of sample, i.e., highlands impact breccia, for example. Y 793274 and QUE 94281 were found on opposite sides of the antarctic continent but are site-paired with each other: they have times of ejection from the moon whose estimated ranges overlap; therefore they *could* have been ejected from the moon in the same impact into a mare. They would be samples from only one point on the lunar surface. Note that all specimens that are fall-paired are automatically site-paired.

HOW MANY IMPACT SITES DO THE LUNAR METEORITES REPRESENT: FALL-PAIRING AND SITE-PAIRING

Because lunar meteorites represent random samples of the surface of the moon, it is important to know how many actual sites on the moon are represented by the 35 lunar meteorites we now have in our collections. Obviously, if all the lunar meteorites we have found came from the same point on the lunar surface they are not telling us much about anything but that single point on the moon. Conversely, if they all came from different sites our information gain from these samples would be greater.

Unfortunately, fall-pairing of lunar meteorites is difficult because the criteria we use for pairing – observed meteorite falls – are lacking. For example, none of them are observed falls, and our estimates of how long ago each of them fell is subject to great uncertainty. Given the existing rarity of lunar meteorites on the earth's surface, finding two in close proximity does suggest that they may be fall-paired. Finding two antarctic specimens far apart, however, does not prove that they are not paired, because two fragments of the same specimen that fell relatively close together may have been transported great distances after fall by diverging ice flows. Another difficulty is that most rocks on the lunar surface are impact breccias composed of bits and pieces of preexisting rocks. These bits may have been thrown great distances across the lunar surface by major impacts and then mixed and churned by lesser impacts before being welded into one highly inhomogeneous rock. Because they are so heterogeneous, it can become very difficult to say that because two lunar meteorites look different, they are *not* fall-paired.

Site-pairing on the moon presents its own set of problems. It is possible, for example, that a single large crater-forming impact on the moon, which, after all, represents only a point on the lunar surface, might inject fragments into quite different orbits about the sun. Lunar fragments from the same impact but in different orbits could collide with the earth hundreds of thousands of years apart in time, and tens of thousands of kilometers apart on the earth's surface; yet both would represent a sampling of the same point on the moon. With this in mind, it may be worth meditating on the suggested site-pairing of Northwest Africa 032 with four lunar meteorites found in Antarctica.

There are two possible criteria we can use to guess whether or not two lunar meteorites are site-paired – i.e., come from the same impact site on the moon. (1) If the evidence suggests that they are fall-paired, it also suggests that they came from a single rock that broke apart when it encountered the earth. That rock could not have been ejected from two sites on the moon, so the two specimens are automatically site-paired. (2) We can usually estimate the length of time a rock has spent in space after ejection from its parent and before fall to the earth. We can also estimate the length of time it has been since it fell to the earth's surface. Within the limits imposed by our estimates, the sum of these two periods gives the age of ejection from the moon. If two lunar samples have the same ages of ejection from the moon, we can be reasonably sure they came from the same impact event, and are site-paired.

The problem here is that cosmochemists can measure exposure ages in space and residence times on the earth only within rather wide limits, so these numbers are not as precise as we would like. Since we can only know these numbers within certain ranges, two lunar meteorites whose ages of ejection have overlapping ranges *could have come* from the same impact event. Examples of this are Northwest Africa 032 and Yamato 793274, which have ages of ejection of 0.04– 0.12 and 0.02–0.05 million years, respectively. Their ranges overlap by 10,000 years, so we cannot say that these two specimens did not come from the same impact event on the moon, even though we may privately doubt that they did. If one had been classified as a highland sample and the other as a mare sample, we could make a case for separate origins, but they are both mare samples, so we do not have that option – we must assume they may be site-paired.

Among the lunar meteorites in the upper part of Table 7.1, if we count each fall-paired group as one specimen, plus the single meteorites that have not been fall-paired, we can estimate a maximum number of 15 sites that have been sampled. If we accept all the site-pairing suggestions, however, we have sampled a minimum number of eight sites. *We can then say that the first 20 specimens listed in Table 7.1 have sampled between 8 and 15 different spots on the moon.*

If we assume all the site-pairings are correct, we have sampled eight sites scattered randomly over the surface of the moon. Six of the sites are in the highlands and two are in the maria. Possibly by coincidence, this is close to the ratio of highlands to maria on the moon: 75% highlands impact sites, as compared to 83% highlands area on the moon. If we use the fall-pairing criterion, according to which we have sampled as many as 15 sites, only 60% were in the highlands.

Waiting for further information are the 15 additional lunar meteorites listed in the lower half of Table 7.1. When their ages of ejection from the moon are known, and the fall-pairing status of each has been determined, we will have larger minimum *and* maximum estimates of the total number of lunar impact sites that have been sampled.

HOW DO LUNAR METEORITES FIT INTO THE OVERALL PICTURE?

During the past five or six years the Clementine and Lunar Prospector orbiters have sent us remote sensing data on the abundances of certain elements over large areas of the lunar surface. The Clementine measurements of iron oxide concentrations over the lunar highlands peak at about 4.5%, which is in excellent agreement with the six or seven lunar highlands meteorites. The lunar meteorites tell us that the Clementine calibration is probably sound, and the Clementine results, taken over vast areas of the moon, confirm for us that the lunar meteorites probably are representative. Lunar Prospector results indicate that the thorium concentration of typical highlands material is lower than that obtained from earlier gamma-ray experiments. The lunar meteorites confirm that the thorium concentration of typical highlands material is low. Because we can see such consistent results between the few elements measured in common by these two methods, we can gain confidence that the lunar meteorites give us the best possible estimate of the total composition of the typical highlands crust. Also, while the Apollo samples all came from the lunar Nearside, many of the lunar meteorites must come from the lunar Farside.

From the foregoing we can say that the lunar meteorites have added to our scientific knowledge of the moon in two ways: (1) they have provided additional lunar samples that we had not had before, whose analysis has extended the known ranges of composition of certain suites of lunar igneous rocks; and (2) they have allowed us to make general statements about the overall geochemistry of the moon with more confidence than we had before because they are samples selected randomly from the entire surface of the moon. Following is a list of some of our ideas before lunar meteorites and how study of the lunar meteorites has affected them.

EXTRALUNAR MATERIAL STRIKING THE LUNAR SURFACE

Before

The moon has received a constant rain of micrometeorites over time, and these tiny impacts have aided the lithification process. These objects also have added their compositions to the upper centimeter or so of the lunar surface at detectable concentration levels. The common surface materials (regolith breccias) collected on the moon by astronauts contain the distinctive low nickel/iridium (Ni/Ir) and gold/iridium (Au/Ir) ratios of chondrites, which are the most common types of stony meteorite. Since the surface *rocks* of the moon are not chondritic, and lunar surface rocks contain very little Au, Ni and Ir, these ratios apparently reflect the composition of most micrometeorites, which then must be chondritic. By contrast, ancient impact-melt breccias that resulted from basin-forming impacts around 3.9 billion years ago have higher Ni/Ir and Au/Ir ratios than chondrites by a factor of the order of 2. These are typical of the impact products of iron meteorites. We can infer from this that while most micrometeorites have compositions identical to the most common types of stony meteorites (the chondrites), iron meteorite impacts formed the basins.

The Apollo 14 and 16 missions provided the Apollo samples collected closest to the lunar highlands. In the Apollo 14 and 16 collections, which presumably were more representative of lunar highland rocks, the regolith breccias have higher Ni/Ir and Au/Ir ratios than those found in chondrites, but lower than is characteristic of iron meteorites. An open question, then, was, "Are these ratios typical of the lunar highlands?"

After

Those lunar meteorites that appear to be regolith breccias from the lunar highlands do not have the higher Ni/Ir and Au/Ir ratios of the

Apollo 14 and 16 lunar highland regolith breccias. This suggests that the typical lunar regolith breccia has chondritic Ni/Ir and Au/Ir signatures, moonwide. Regolith breccias from the Apollo 14 and 16 missions are now believed to be peculiarities of part of the lunar Nearside, due probably to the Imbrium impactor, which would have to have been an iron, not a chondrite.

THE MAGMA OCEAN

Before

Most workers were convinced that the initial igneous rock-forming event on the moon derived from a moonwide magma ocean. As the magma ocean cooled, the first minerals to form were anorthite (a high-calcium, high-aluminum silicate that crystallizes at very high temperatures) and olivine (a silicate that crystallizes initially with high magnesium contents at high temperatures). The olivine, being denser than the magma, settled downward while the anorthite, with a lower density than the magma, floated upward. This fractionation process created a sort of thick mush at the surface of the moon, made up of anorthite crystals floating in variable amounts of still-liquid magma from which the high-temperature, magnesium-rich olivine crystals had been removed. When the magma cooled further, the remaining liquid crystallized to form the somewhat more iron-rich minerals that typically crystallize at lower temperatures. This process created a very anorthite-rich crust over the surface of the entire moon.

After

Do the lunar meteorites support this model? The short answer is "yes, they do." The high content of Al_2O_3 in most of the lunar meteorites supports a model in which anorthite has been concentrated to a higher degree than would ever be possible if fractionation had not occurred. Not only did high-temperature olivines fractionate downward in the magma from which the anorthite-rich rocks formed, but any co-forming high-Ca pyroxenes also settled out. The only way for such an extreme fractionation to become efficient is in a magma body of very great size – essentially, a moonwide magma ocean that crystallized to form the original crust of the moon.

THE COMPOSITION OF THE LUNAR HIGHLANDS

Before

An early plot of the compositions of lunar highland rocks derived from the Apollo highlands samples (Figure 7.3) showed two distinct compositional series: one is the linear, so-called *ferroan anorthositic suite* in which the dominant mineral is anorthite, which characteristically crystallizes at very high temperatures. The high temperature magnesium-rich minerals that would be expected to crystallize at similarly high temperatures, however, are not present in this series. Instead, the magnesium–iron (Mg/Fe) silicates that do occur in association with anorthite have lower Mg/Fe ratios that are characteristic of lower crystallization temperatures. Lower Mg/Fe means richer in Fe; hence the name "ferroan anorthosites." These, and other anorthite-rich rocks are postulated to have formed a moonwide crust over the magma ocean, as discussed above. The second (*Mg-suite*) series has compositions we are more comfortable with because they follow trends that we would expect from our terrestrial experience with differentiated magma bodies. The Mg-suite series may represent a group of later intrusions into the anorthositic crust. In 1980, these compositional trends appeared to be completely separate, as shown on the plot. The closest terrestrial analogy to these two trends occurs in the Stillwater igneous complex of Montana, an ancient igneous intrusion more than 5 km thick, but in that case the two series intersect near their high-Mg ends.

After

By 1991, the range of compositions in the ferroan anorthosite series had been extended to both higher and lower Mg/Fe ratios, partly by addition of new analyses of Apollo mission samples but also, in an important way, by data from lunar meteorites (Figure 7.4). Currently, the extremes of the ferroan anorthosite trend are anchored in the

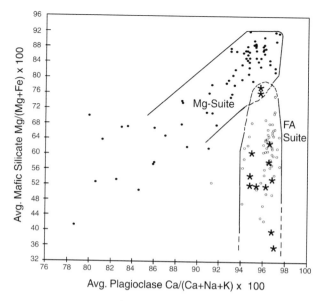

FIGURE 7.4 This diagram is a later version of Figure 7.3. Data points from lunar meteorites are indicated by stars. Notice that with the addition of more analyses of rocks and minerals from Apollo Program lunar samples, the composition range of the ferroan anorthosite family of lunar rocks and minerals has been extended. The extremes, at both ends, have been supplied by lunar meteorites. (Adapted from Warren, P. and G. Kallemeyn (1991) The MacAlpine Hills lunar meteorite and implications of the lunar meteorites collectively for the composition and origin of the Moon, *Geochimica et Cosmochimica Acta* **55**, 3123–38.)

compositions of pristine clasts from lunar meteorites; now the two series appear at least to touch, if not to overlap. So the analogy with the chemical variations in the Stillwater complex perhaps is now a little stronger due to data from lunar meteorites. The immediate significance of this, however, is uncertain, since the origin of the anorthosite bodies in the Stillwater also is still somewhat obscure.

POTASSIUM, THORIUM AND URANIUM IN THE MOON
Before

Heat is being generated in planets today by radioactive decay of certain isotopes of potassium (K), thorium (Th) and uranium (U). Based on the Apollo samples and analyses of the Russian Luna samples, it had been concluded that, compared to the earth's mantle, the moon is greatly enriched in these radioactive elements. But the Apollo and Luna samples were collected in or near lunar Nearside mare regions, and it could be suspected that these areas would not be representative of the highlands. Those samples collected by the Apollo 16 astronauts were thought to most closely represent highlands materials, and these did indeed contain lower concentrations of these elements. Concentrations of K, Th and U in lunar surface materials can be estimated by satellite measurement of the gamma radiation they produce during radioactive decay. Remote sensing studies of this type had shown progressively decreasing concentrations in these elements with increasing distance from the lunar Nearside, but the estimated concentrations of these elements for the bulk moon still were several times as great as those we believe to be present in the earth's mantle.

After

Lunar meteorites thought to represent highlands regolith breccias have the lowest levels of Th and U yet detected in lunar rocks and do not show variations that would support the earlier-reported Th + U concentration gradient as a function of distance from the maria. Estimates of K, Th and U in the bulk moon have now been lowered to levels believed to be present also in the earth's mantle.

Before

Regoliths are very porous and provide an insulating layer over a planet's surface that tends to conserve internal heat. Regolith breccias are formed at, or very near, the lunar surface by impacts into regolith. The Apollo mare samples consisted of a very high percentage of regolith breccias. However, only a small fraction of highland regolith breccia samples were collected by the Apollo 16 astronauts. Because the Apollo 16 site was thought to come closest, among all the Apollo sites, to representing the lunar highlands, this led to speculation among some workers that the regolith is much thinner in the lunar highlands than in the maria. This, in turn, lent some credence to the then-current high estimates of the K, Th and U concentrations in the moon because a less-well-insulated highlands surface would allow the heat generated by these elements to escape more efficiently.

After

The low incidence of regolith breccia samples in the Apollo 16 lunar highlands collection may have resulted from a conscious effort on the part of the astronauts to seek out and collect rocks that appeared to be pristine; thus producing a biased sample. The lunar highlands meteorites, which presumably provide an unbiased sample, consist almost completely of regolith breccias, confirming what most researchers really believed: that the entire moon is well insulated by regolith. Thus, it would require lower concentrations of K, Th and U (the heat-generating elements) to maintain the observed thermal emission at the lunar surface. This conclusion is consistent with the conclusion listed earlier that the moon, in general, has lower concentrations of these elements than had been believed.

IS THERE A THIRD GREAT PETROLOGICAL PROVINCE ON THE MOON?

Before

We tend to think about the moon as having two quite different zones: (1) the lunar highlands, which formed in response to fractional crystallization in a moonwide magma ocean; and (2) the nearside maria that have a high abundance of rocks formed from lava that flowed out at the surface to fill the impact basins. This lava probably was derived from fractional melting of very deep-seated rocks. In general, however, after comparing Apollo lunar samples and lunar meteorites, we have begun to appreciate the fact that the lunar nearside contains chemically exotic compositions, compared to the overall lunar surface. A group that includes Larry Haskin, Randy Korotev and others at Washington University have concluded that rocks enriched in incompatible elements were produced at only one site on the moon: an area on the side toward Earth, which includes Oceanus Procellarum and the Imbrium Basin, but does not include some of the other maria.

Incompatible elements find it difficult to fit into the crystal structures of the higher temperature minerals and therefore tend to become concentrated in the last magma to crystallize. This group of elements includes the rare-earth elements (REE), phosphorus, K, Th and U. Since the latter three elements generate heat, their concentration would tend to prolong the molten stage of any accumulation of magma in which they were concentrated. In the view of the Washington University group, the Imbrium impact occurred near the northeast edge of this unique site on the moon. When impact by an asteroid produced the Imbrium Basin it directly tapped this "hot spot," which may still have been partly molten, showering the immediate surroundings with impact products that were relatively enriched in incompatible elements. All of the Apollo program's "lunar highland sites" are close enough to this impact basin to have been covered with impact debris to a depth of hundreds of meters. The entire lunar surface probably received significant contributions of impact breccia of this possibly unique composition, but it decreased with distance from Mare Imbrium. Lunar orbiters would record high concentrations of K, Th and U on the lunar nearside, but levels of these elements would decrease away from the nearside in response to the decreasing incidence of Imbrium impact products deposited on the surface. Even so, the moonwide distribution of Imbrium materials led to erroneously high estimates of K, Th and U in the moon.

After

To date, lunar meteorites neither support nor weaken the idea of a third petrological province on the moon. They are compatible, however, with the notion that the lunar highlands have Earth mantle-like concentrations of incompatible elements, but in some regions have received variable amounts of basin-impact debris that is rich in incompatible elements.

WHY ARE THERE NO IMPACT MELTS THAT DATE BACK TO THE MOON'S ORIGIN?

Before

Impacts on the moon that are intense enough to form craters from about 20 to 1200 km in diameter can reasonably be described as major. Such impacts are smaller than the basin-forming impacts that produced the maria but large enough to produce impact melt sheets. These melt sheets are massive enough to allow slow cooling. Under such conditions, included gases such as previously existing radiogenic ^{40}A can escape and networks of tiny crystals can form. The resulting rocks are newly crystallized igneous rocks. The $^{40}K/^{40}A$ clock has been reset, and if we can determine the age of such a rock that has been formed from impact melt, we have measured the date at which a major impact occurred on the moon.

Many of the lunar samples brought to Earth by the Apollo astronauts were breccias, and these contained inclusions identified as impact melt fragments. The ages of these impact melts were quite ancient, but none were older than about 4 billion years. So we had no record of major impacts on the moon during the first 600 million years, or so, of its history. This might not be surprising, because the Apollo samples have been collected from a very small area on the lunar nearside, in a region of the moon that suffered basin-forming impacts and subsequent outpourings of lava (the maria) that could have erased all earlier records on the nearside. Still, there were many workers who interpreted the data as an indication that the moon had suffered a major bombardment around 4 billion years ago that erased earlier records all over the moon. The lunar samples have now given us a chance to test this hypothesis.

After

The lunar highlands are more ancient than the maria, and have not been overprinted by more recent outpourings of lava, so impact melt fragments in lunar highlands meteorites should contain a record of any early bombardment of the moon. Barbara Cohen at the University of Tennessee, and colleagues Tim Swindle and David Kring at the University of Arizona have determined the ages of 31 impact melt fragments in four lunar highlands meteorites. These ages describe a minimum of seven major impacts that occurred in highland regions of the moon, and none are older than 4 billion years. Their conclusion is that the lunar highlands also contain no record of major impacts during the first 600 million years of the moon's existence. These findings greatly strengthen the inference that the moon suffered a cataclysmic bombardment about 600 million years after the moon and planets formed. Were the first 600 million years of the moon's existence very quiet years, or did the cataclysmic bombardment erase all previous major impact records that may have existed? We do not know.

An interesting sidelight, as Cohen and her colleagues point out, is that it would be surprising if only the moon had been affected by this bombardment. They estimate more than 1700 major impact events occurred over the entire lunar surface about 4 billion years ago, and this would translate to more than 17 000 striking its near neighbor, the earth, at the same time. It also seems quite likely that at least the entire inner solar system was affected. They infer that the effects on the earth would have included the production of large amounts of ejecta, a temporary injection of hot silicate vapor into the atmosphere, and the boiling away of large quantities of surface water. Perhaps coincidentally, the earliest isotopic evidence of life on earth is around 3.9 billion years old. Was all earlier life destroyed by this bombardment, only to arise again, or did a swarm of impactors deliver precursor molecules and create hot, high-energy conditions conducive to the original beginning of life on Earth?

Many questions about the origin and evolution of the moon cannot be answered by either the Apollo samples or the lunar meteorites we have collected so far. Remaining problems include the heat source that produced the moonwide magma ocean, the cause of the cataclysmic bombardment of the inner solar system 4 billion years ago, and the origin of the moon itself. Perhaps additional lunar samples might render clues to some of these questions, but in light of our shamefully shortsighted decision to discontinue lunar landings and associated field research on the moon we cannot expect help from that source. Our recourse is continuing to search for, and find, more lunar meteorites.

SUGGESTED READING

Cameron, A.G.W. (2001) From interstellar gas to the earth-moon system. *Meteoritics and Planetary Science* **36,** 9–22.

Cohen, B.A., Swindle, T.D. and Kring, D.A. (2000) Support for the Lunar Cataclysm Hypothesis from lunar Meteorite Impact Melt Ages. *Science* **290,** 1754–56.

Hartmann, W.K., Phillips, R.J. and Taylor, G.J. (Eds.) (1986) *Origin of the Moon.* Houston, TX: Lunar and Planetary Institute.

Heiken, G.H., Vaniman, D.T. and French, B.M. (Eds.) (1991) *Lunar Sourcebook.* New York: Cambridge University Press.

Papike, J.J., Ryder, G. and Shearer, C.K. (1998) Lunar samples, chapter 5, *Planetary Materials,* ed. J.J. Papike, *Reviews in Mineralogy,* vol. 36. Washington, D.C.: Mineralogical Society of America.

8 How, and where, in the solar system . . . ?

INTRODUCTION

Around 4.57 billion years ago, our part of the Galaxy was approaching a cusp in time and space that, once passed, would see the beginning of an irreversible process of star and planet formation. Our solar system would result. Just before it happened, our cloud of gas and dust had a past but no future – it wasn't quite dense enough on its own to begin gravitational contractions that would result in the birth of a star and associated planets. With no external stimulus it probably would just remain a cloud – formless, highly diffuse and without apparent purpose. But in a very intimate sense it was *our cloud* – we were all there, represented unknowingly by the atoms of which we are today composed.

A cusp is a point defined by the tangential convergence of two curves. The time line of our cloud was converging with the time line of a nearby giant star that had become unstable and was set to collapse inward with unimaginable intensity. This would initiate a supernova and splash part of itself out into space in an ejaculation of cosmic violence. Part of this giant splash was directed toward us (to be). The first signal of the nearby supernova was a flash of electromagnetic radiation, of which the part we call visible light is a tiny segment, washing into and through our cloud; perhaps for the first time illuminating its murky interior for no one to see. Radiation pressure exerted a weak force on the matter in the cloud, imposing small parallel force components on the more or less random movements of its atoms and dust particles. Gas molecules and the tiniest solid particles were most affected by this, and the cloud began to densify a little bit as these common components of motion, superimposed on their former random motions, caused a barely perceptible parallel drift in a direction away from the supernova. Some time later, our cloud was impacted by waves of plasma from the supernova, in the form of electrons, protons, helium nuclei and, more rarely, nuclei of all the elements and their isotopes. Some of these particles were slowed enough by collisions to take up residence in our cloud, and there is a good chance that each of us contains in our makeup some of these supernova-born atoms[1].

Compared to the radiation pressure from the initial flash that announced the supernova, collisions with the heavier particles in the second wave were like hammer blows, all delivered in the same direction, and the side of our cloud that was facing the supernova began retreating. As this retreat continued, more and more mass was concentrated into a smaller volume until some of the lumpier regions within the cloud began to contract gravitationally to form growing accumulations of matter. One region in the cloud may have attained critical density and begun attracting mass to itself before the others; it grew very large and became the sun. Somewhat later other, relatively dense regions of the cloud produced other accumulation centers – soon to become planets or lesser bodies. So wherever the density gradients were high enough, those accumulation centers began augmenting their mass by gravitationally attracting matter from the surrounding cloud which, by now, was revolving about the largest accumulation – soon to become the sun. The formless *primordial cloud* had become the disc-shaped *solar nebula*, containing gas, dust grains and possibly larger bodies.

Very difficult to explain in this abbreviated, *and highly speculative*, history is the existence of calcium–aluminum-rich grains that sometimes are enriched in oxygen isotope ^{16}O. Believed to have formed at a very early stage within the primordial cloud or the solar nebula, their mineralogy indicates that they formed at very high

---

[1] Actually, there is nothing really special about this. All the elements upward from helium in atomic mass were born in stars, and most of those elements found in the primordial cloud represent debris from earlier supernova explosions and the expanding envelopes of matter around red giant stars that took up residence in our end of the universe. This is nature's way of providing the raw materials to form new stars and planets (and people).

temperatures, and their oxygen-isotope ratios may reflect some influence from the nearby supernova's expanding gas cloud. ^{16}O has the most stable nucleus of the oxygen isotope, and is the specific isotopes we would expect to be able to survive the holocaust of a supernova event.

Within the solar nebula, the smaller mass accumulations were rotating, but also revolving about the central mass. Much later, when much of the mass in our cloud had concentrated itself either into the birthing sun, the nascent planets (not as yet discrete planetary bodies), or into the central plane of the solar system, a third and final assault arrived from the dying supernova. This took the form of much slower-moving dust grains composed of refractory compounds that had formed by condensation at high temperatures in the relatively cooler outer reaches of the supernova's expanding gas cloud. Most such grains were silicon carbide and graphite, with a small admixture of other carbides that form at extremely high temperatures. These particles also contained rare gases such as neon and xenon with highly anomalous isotopic ratios that would be expected to occur only as products of a supernova event. Because of their isotopic anomalies they can be identified in some meteorites. As they can be embedded in meteorites, it seems that at least some meteorite parent bodies were forming even at this early stage in the birth of our solar system. This brief history has led us back to meteorites, of which the most prolific current source is Antarctica.

A SIMPLE CLASSIFICATION OF METEORITES, FOR PEOPLE WHO DON'T KNOW IT ALREADY

Depending on their overall compositions, meteorites are of three types: **Irons**, **Stones**, or **Stony Irons.**

This is a very useful classification because it is simple, and it tells us something in words with which we are familiar. "Irons," however, are composed not just of metallic iron, they have varying amounts of *alloyed* nickel and cobalt, and trace concentrations of many other elements that act chemically like metallic iron. Irons also can have sulfides, phosphides, carbides and other compounds that, insisting upon chemical exclusivity, are found as nodules segregated within the main mass. Some irons also have silicate inclusions. The degrees to which these alloyed elements and associated inclusions are present, and their relative proportions, have motivated metallurgists and chemists to suggest many subclasses of the irons. These are called groups. We assume that each distinct group is probably characteristic of a different parent body. This may, or may not, be a valid assumption.

Stony irons are composed of significant proportions by volume of metallic iron (as in the irons) and silicates (as in rocks and minerals that make up the earth's mantle).

Stony meteorites are generally similar in composition to our model for mantle rocks of the earth, but most of them also contain flecks of metallic nickel-iron which we would not expect to see in terrestrial mantle rocks. While their compositions are similar to those we imagine mantle rocks to have, they do not look like what we would visualize mantle rocks to look like, because they have not been subjected to the high pressures and temperatures present in the earth's mantle.

While stony irons have only a few subclasses, meteoritic irons and stones have many. It is important to understand some of the differences between the subclasses (groups) of stony meteorites, for what they can tell us about the primordial cloud, the solar nebula *and* the evolution of the parent bodies of meteorites. To begin with, there are two major divisions among the stony meteorites: chondrites and achondrites.

THE SOCIOLOGY OF CHONDRITES
Mixed neighborhoods

Chondrites would remind a sociologist of the "melting pot" effect, often cited as a significant process in the evolution of modern society in America, in which immigrants from many parts of the world brought highly diverse backgrounds to a new environment in which they were forced to interact with persons from other backgrounds.

Over several generations stresses in the new environment should have a homogenizing effect on members of these different cultures, resulting in less and less diversity as larger and larger bits of the old cultures are erased, modified, or adopted into the new environment. While theory has predicted the melting pot effect with less than complete success in human populations, it performs perfectly in chemical systems such as those of chondrites.

When examined closely, nine out of 10 meteorites are chondrites. Each chondrite seems to have been, in its primitive state, a random accumulation of bits and pieces that must have come from a large number of different neighborhoods, all thrown together and expected to make friends. Coexistence between grains in chondrites, however, had potential problems: some bits and pieces had come from hot and some from cold environments, and neither group can be thermodynamically ecstatic over the new neighbors. Some individuals from hot sources were molten droplets that had cooled more or less rapidly while still preserving a spheroidal form. These interesting spheroids are called chondrules, and chondrites take their name from the presence of chondrules. Others from hot environments were simple mineral fragments that may have crystallized as parts of chondrules, or as components of cooling magmas. They came to the new environment with a history of broken homes. Other residents, the ones from the coldest environments, arrived as complex hydrocarbons, some of which cannot exist at temperatures much above 200 °C.

Different chondrite neighborhoods have been subjected to different degrees of stress, in the form of thermal energy. As heat builds in a primitive planetary pile, the complicated hierarchies of the hydrocarbons are the first to break up, sending derivative sons and daughters diffusing out into the neighborhood, where some of them are assimilated. With increasing thermal stress, close neighbors, in a sintering process, begin to identify more closely with each other than with more distant neighbors and at successively higher stress levels more and more early barriers are broken down and intermarriage occurs. These chemical reactions in response to heating completely change the character of the original families and the children retain ever fewer memories of the old heritage. If the process is carried to completion the neighborhood is completely homogenized. Meteoriticists understand the metamorphic sociology of chondrites. They interview in the unstressed neighborhoods to hear the ancient songs of the solar nebula and in the more highly metamorphosed associations to get the later news.

Thermal stress, and metamorphic grades among the ordinary chondrites

Metamorphic grade is a judgment call by petrologists, who estimate how much thermal-induced solid-state recrystallization and homogenization of mineral grains has occurred, along with the degree to which the chondrules are recrystallized and have reacted with the groundmass to blur their original outlines. Petrologists originally classified chondrites, i.e., those meteorites that contain chondrules, into six proposed metamorphic grades, in which completely primitive and unaltered samples were of grade 1 and completely homogenized specimens were of grade 6.

Remember that about 90% of the meteorites that fall are chondrites. Among the chondrites, most are closely related in bulk composition and mineralogy; these most common chondrites are called ordinary chondrites. Ordinary chondrites show metamorphic grades 3–6. Recently, in an initial way through the work of Gary Huss at the University of New Mexico and, more completely, in the thermoluminescence studies by Derek Sears at the University of Arkansas, we have realized that grade 3 in the metamorphic series of ordinary chondrites brackets a number of identifiable substages involving progressive thermal metamorphism. The metamorphic effects realized exclusively within grade 3 may be at least as significant as those occurring in the succeeding grades 4–6. As a result, ordinary chondrite grade 3 is now subdivided into grades 3.0–3.9.

In addition to being subdivided according to metamorphic grade, ordinary chondrites occur in three discrete groups that probably represent different parent bodies. These groups are called H (high), L (low) and LL (really low), according to how much total iron (Fe^o, Fe^{2+} and Fe^{3+}) they contain. LL is different from L in having not only low total iron, but proportionately lower metallic iron (Fe^o), as well. So, to summarize, including groups and metamorphic grades, ordinary chondrites include all of the following: (H 3.0–3.9, 4, 5, 6); (L 3.0–3.9, 4, 5, 6); and (LL 3.0–3.9, 4, 5, 6)[2].

Other, rarer, chondrite groups

Aside from the ordinary chondrites, the current classification recognizes: enstatite chondrites (E), Rumuruti chondrites (R) and carbonaceous chondrites (C). Enstatite chondrites are so named because they are richer than all other chondrites in the mineral enstatite ($MgSiO_3$). There are two groups of enstatite chondrites, designated in shorthand as EH and EL, where H and L refer to high and low levels of total iron, as in the ordinary chondrites. R (or Rumuruti) chondrites, named after an observed fall that occurred on January 28, 1934 at Rumuruti, Kenya, has very recently been identified as a separate group. Carbonaceous chondrites consist of six groups: CI, CM, CO, CV, CK and CR (no relation to Rumuruti), all named after observed falls and all exceptionally rich in carbon. Twelve chondrite groups, each thought to have come from a different parent body, located at a different, but unknown, distance from the sun, can be summarized as follows: EH, EL, H, L, LL, R, CI, CM, CO, CV, CK and CR. There are also some probable new groups of carbonaceous chondrites on the horizon.

Well, classifications tend to be rather tedious, and this one is no exception. However, the exceptional reader, i.e., that one who is still alert, will be asking, "What about metamorphic grades 1 and 2 – the ones that were going to sing to us of the primordial cloud?" This is a good question with a somewhat complicated answer.

---

[2] This classification contains an assumption, in the sense that no H 3.0–3.1 specimens have yet been found, anywhere.

WHAT IS A PRIMITIVE METEORITE?

We find metamorphic grades 1 and 2 among the carbonaceous chondrites, but not among the other types. Most of the early chemical analyses of carbonaceous chondrites were carried out by Birger Wiik, who proposed to subdivide them into three types labeled as Roman numerals I, II and III. Wiik's Type III carbonaceous chondrites correspond to what a petrologist would call metamorphic grade 3, and there was an initial tendency to regard Types I and II as equivalent to metamorphic grades 1 and 2, respectively. This seemed logical because Types I and II have fewer high temperature components and more volatile compounds, including water. They also seemed to fit in nicely with the other chondrites because, while other chondrite groups had no members lower than metamorphic grade 3, carbonaceous chondrites had no members with metamorphic grades higher than 3, and so they seemed to complete the metamorphic series. One exception to this is that we now have a group of carbonaceous chondrites, the CKs, that consist of metamorphic grades 4–6. Another, more serious problem, however, is that carbonaceous chondrites of Wiik's types I and II seem to have suffered metamorphism of a different sort – low temperature aqueous alteration which, by hydration, has apparently changed preexisting higher-temperature silicate minerals to low-temperature clays. Type II carbonaceous chondrites are a mixture of hydrothermally altered silicates, but with some chondrules and isolated high-temperature mineral grains still surviving, and Type I contain principally hydrothermally altered silicates, with rare high-temperature mineral grains that may be all that is left of earlier chondrules. So we must now think of grades 3–6 as demonstrating progressive *thermal* metamorphism and, going in the other direction, grades 3–1 as having suffered progressive *aqueous* metamorphism.

The present view is that the most primitive chondrites are of metamorphic grade 3.0, with those 3s of higher grade having suffered progressive thermal metamorphism and those of lower grade (now called 2 and 1) having suffered progressive low-temperature aqueous alteration, with grade 1 being the most altered. The confusion does not end here, however, because Wiik's Type I carbonaceous chondrites are closest *in composition* to the composition of the sun, ignoring the sun's tremendous excess of hydrogen and helium. Since the sun contains most of the mass in the solar system, deviations from the average composition of the sun indicate some kind of compositional evolution. Type I carbonaceous chondrites therefore could be thought of as compositionally primitive but petrographically evolved, while grade 3.0 chondrites could be thought of as metamorphically primitive but compositionally evolved.

Types I, II and III are not used anymore. Type I is now called CI chondrite, where, instead of the Roman numeral I, "I" now stands for Ivuna, a meteorite that fell in 1938 near Ivuna, Tanzania. Type IIs are called either CM chondrites after Mighei, an alternate name for a Type II carbonaceous chondrite that fell near Olviopol, Russian Ukraine, in 1889, or CR chondrites, named after Renazzo, a stone that fell near Ferrara, Italy in 1824. So far, all known examples of CMs and CRs are of metamorphic grade 2. Because of chemical and isotopic differences that suggested they may not all be from the same parent body, Type III carbonaceous chondrites also received two different group names: CO, after Ornans, an alternate name for a fall in 1868 near Doubs, France and CV, after Vigarano, the synonym for a fall in 1910 near Ferrara, Italy. The only known members of these two groups are of metamorphic grade 3, but presumably grades 4–6 are possible, so they are commonly referred to as CO3 and CV3. CM chondrites are cited as CM2s, although there may be a few CM1s that have been hydrothermally altered to a greater extent than is typical of alteration grade 2. All known CI chondrites are of alteration grade 1 and all known CR chondrites are of grade 2, so seeing "CI1" or "CR2" in the literature is unusual.

It is an interesting coincidence that the type specimens of two rare meteorite groups, CVs and CRs, fell about 16 km from each other on the earth's surface. Of course, they fell 86 years apart in time; otherwise I might be tempted to retire to Ferrara in hopes of seeing the fall of another new type specimen. Even so, it is now 90 years after the fall of Vigarano, so maybe another one is due.

CK carbonaceous chondrites are named after the type specimen Karoonda, which fell in Buccleuch County, South Australia. These metamorphic grade 4–6 specimens are the first carbonaceous chondrites to show the effects of thermal metamorphism beyond grade 3.

KERR GRANT DISCOVERS A METEORITE

It is perhaps a strange coincidence that in 1954, I was given a few small pieces of Karoonda by its discoverer, Sir Kerr Grant, and 28 years later was associated with the antarctic field project that recovered the eight specimens that, with Karoonda, were sufficient to characterize a new carbonaceous chondrite parent body – the CKs. In 1954, I had received my BS in geology at the University of New Mexico. I had then worked for a while as a seismic computer for the Superior Oil Co. of California (before we had electronic computers) and then received a Fulbright scholarship to the University of South Australia, in Adelaide. Kerr Grant was Emeritus Professor of physics at the university. He was famously absent-minded, and tremendously popular among the students. Stories of his absentmindedness are rife. I became acquainted with him when I received permission to sit in with a small discussion group that he conducted every year. During that time I elicited his first-hand version of the circumstances surrounding the recovery of the Karoonda meteorite. I do not believe this story has been recorded anywhere, and the events as they occurred are sufficiently unusual as to warrant making it a matter of record.

The meteorite fell at 10.53 p.m. on November 25, 1930. The fall was observed over a wide area. About a week later, after receiving numerous reports from witnesses to the fall, Kerr Grant and G. F. Dodwell, Government Astronomer for South Australia, decided to mount a small expedition to try to recover the meteorite. They interviewed a number of witnesses in person, driving from place to place and camping at night in a tent. On the morning of December 10, Kerr Grant awoke quite early. His companion was still asleep, and rather than waking him, he decided to walk over to a nearby farmhouse to see if anyone there had witnessed the fall. The lady of the house was in the yard, hanging clothes out to dry. He asked her if she had seen the meteorite. She said no, she hadn't, but her daughter had seen it hit the ground. He became quite excited at this news, and demanded to talk to the daughter immediately. The daughter was produced and, standing exactly where she had been standing when she witnessed the fall, pointed to a tree on the horizon. She said she saw the rock strike the ground in an open field directly in line with the tree. She must have had excellent vision, as the fall had occurred close to 11 p.m.! He thanked her and hastened off toward the indicated tree. After walking some distance, however, his attention wandered and he began thinking about something else; his absentmindedness had kicked in. He was brought back to an awareness of his surroundings when he stepped into a hole in the ground. Crouching with one foot in the hole and his other leg bent double, he was preparing to extricate himself when he noticed a scattering of strange-looking rocks. He was down close to them, so they were quite noticeable. Picking one up, he decided that it might be a meteorite, and correctly surmised that his foot was resting directly on the main mass! That is how the Karoonda meteorite was found, as related by Sir Kerr Grant to a young Fulbright student, in 1954.

RECENT DEVELOPMENTS

A new carbonaceous chondrite group, CH chondrites, has been described by Addi Bischoff, of the Institut für Planetologie, Münster, Germany and a group of colleagues at his institute and at Max Planck-Institut für Chemie in Mainz, Germany. They may be related in some ways to CR chondrites but are different enough to justify not including them in that group. "H" in the designation CH refers to their uniquely high total iron content among carbonaceous chondrites, and is used in an analogous manner to EH, for high-iron enstatite chondrites, and H-type ordinary chondrites, where H also stands for high iron contents. This new group, along with the CR chondrites, considerably extend the range of chemically and mineralogically distinct subgroups of carbonaceous chondrites.

WAS THE SOLAR NEBULA INHOMOGENEOUS? EVIDENCE FROM THE CHONDRITES

Given our model of the primitive solar nebula as a flattened, but possibly lumpy, cloud of gas and dust revolving about a major central accumulation that was starting to heat up, it is logical to wonder if the disk was homogeneous throughout, or had been fractionated in any way, from center to outer edge. In other words, could there have been intrinsic differences in the reservoirs of matter that were available to each accumulating body in the solar system? As we saw earlier, the least altered meteorites are most likely to retain records of the characteristics of the solar nebula, so it is among the E3, LL3, L3, H3, R3, CM2, CR2, CV3 and CO3 chondrites that we must search for any such evidence.

In an interesting first approach to such a study, Alan Rubin and John Wasson, at the Institute of Geophysics and Planetary Physics, University of California at Los Angeles (UCLA), have pointed to a number of such differences between these groups of least-metamorphosed chondrite classes. They find regular variations across these groups in the following characteristics.

- Oxidation state.
  Relative oxidation state is often expressed in the ratio $FeO/(FeO + MgO)$ (iron oxide + magnesium oxide). This term is called the FeO number, and is frequently written as FeO* or as FeO#.
- Refractory lithophile abundance.
  As used here, refractory lithophiles are elements that prefer to be incorporated into silicate minerals that crystallize at high temperatures.
- Matrix/chondrule modal abundance ratio.
  Modal abundances are determined by point-counting under the microscope. This is a statistical measure of the volumes of chondrules vs. matrix material present in the meteorite.

- Degree of chondrule melting.
  Some chondrules appear to have been completely melted droplets; others show evidence of only partial melting and contain relict, unmelted grains within them.

Listed above are four of the five characteristics examined by Rubin and Wasson (more about the fifth one, later). The interesting result they describe is that the major groupings (and here I have combined some chondrite groups into major groupings), show differences from each other in those characteristics. This suggests that there were differences in the reservoirs of solar nebular material from which each supergroup formed. Even more interesting is that if the supergroups are arranged as follows:

enstatite chondrites – ordinary chondrites – R chondrites – carbonaceous chondrites

then the four characteristics vary in a regular manner from one to the next: FeO*, refractory lithophile abundance, and matrix/chondrule modal abundances all *increase* from left to right and degree of chondrule melting *decreases* from left to right. Assuming that these chondrite groups sampled different regions of the asteroid belt, we can suspect that the solar nebula did indeed vary somewhat in its characteristics, as a function of radial distance from the center of the nebular disc[3].

The fifth characteristic they discussed was oxygen isotopic composition. This is sometimes expressed in terms of $^{17}O/^{16}O‰$ (i.e., parts per thousand) relative to the same ratio in "standard mean ocean water." They describe the oxygen isotopic composition as *increasing* in the following sequence: EH – EL – H – L – LL – R; but then *decreasing* across the following sequence: CR – CM – CV – CO – CK. Note that they break the major groupings down into discrete groups here. For ease of comparison, I used the major groupings for the other

---

[3] In fairness to Rubin and Wasson, I have to say that while I use their data, my speculations are somewhat different and I do not discuss the cause(s) of these variations, which they ascribe to an extreme radial temperature gradient within the solar nebula.

four. If we use only the same major groupings for the oxygen isotopic composition, we still do not see a steady increase or decrease across the major group sequence as listed above. In thinking of possible reasons for this, it occurred to me that the first four characteristics were almost certainly intrinsic to the nebular disk, but that the oxygen isotopic composition could have been imposed from outside the disk. Remember that we find isotopic anomalies in meteorites. Most of these anomalies probably came from the nearby supernova that triggered the initial collapse of our cloud. Remember also that the solar disk may have been somewhat lumpy, so an amount of ^{16}O contributed to a large lump in which a meteorite parent body was forming would have less relative effect than the same amount deposited in a smaller lump, in which an ultimately smaller meteorite parent body was forming. In any case, the oxygen isotopic compositions probably should not be compared to the other four characteristics, but they could conceivably be related to an independent parameter, i.e., the relative thickness of the solar nebula at the point where the parent body formed and by inference, perhaps, to the relative size of the resulting parent body.

WHY IS IT IMPORTANT TO SPEND TIME ON CHONDRITES?

Chondrites have a complicated classification scheme that is annoying to try to learn, but study of this large subtype of stony meteorites is worthwhile because the more primitive ones give us hints about conditions in the solar nebula. Conversely, the specimens of higher metamorphic grade, because they have been thermally metamorphosed during deep burial in their parent bodies, can give us a sense of thermal effects in the mantles of these bodies.

THE ROLE OF ANTARCTICA AS A SOURCE OF CHONDRITES

The antarctic meteorite collection is known to be approaching 30 000 specimens. That number may have been attained by the time this book has been published. Any discussion of an ever-growing accumulation of objects will be out of date by the time the conclusions are heard. Recognition of this fact gave me an excuse to limit my data set to those specimens reported before Jan 1, 2000, as presented in *The Catalogue of Meteorites* (fifth edition). The great advantage to this decision is that time does not have to be spent adding later data to those in the Catalogue, and the data existing in this Catalogue already exist on a CD that accompanies the Catalogue and can be sorted electronically. The results are listed in Table 8.1.

It should be recognized that the list of non-antarctic chondrites includes only distinct falls, plus finds that probably result from distinctly separate falls. A single fall can consist of one specimen or, rarely, hundreds. The antarctic meteorites are all finds. Because hundreds or thousands of specimens can be found on a single large stranding surface, there is no way to know how many separate falls are represented. For purposes of comparison by numbers, all we can say is that the numbers of antarctic meteorite specimens found are greater than the number of actual falls that produced them. It is a good general principle, however, that since meteorites are as rare as they are, the more specimens one can examine, the greater is the chance that one will note some previously unobserved feature in one of them. Such a feature, found in a single specimen, might provide an insight into the conditions of origin of the entire group, or give a clue to the state of the solar nebula at the time the parent bodies were forming; so numbers are important. In comparing the numbers, antarctic meteorites can be seen to have added significantly to some of the rarest of the rare, such as the newly discovered Rumuruti and Kakangari-like groups and the CK, CR2 and CH groups among the carbonaceous chondrites.

ACHONDRITES AND THE HISTORY OF PLANET FORMATION

Achondrites can be igneous rocks crystallized from a magma or, alternatively, metamorphic rocks that have been so highly heated that they have recrystallized extensively at temperatures around those required for partial melting. In parent-body structure studies, the achondrites

Table 8.1. *Comparison of the numbers of non-antarctic and antarctic chondrites, as tabulated in* The Catalogue of Meteorites[a]

Group	Non-antarctic	Antarctic
EH + EL	84	120
H	1648	5307
L	1373	4791
LL	243	805
R	8	11
K	2	1
CI	5	0
CM2	22	138
CO3	46	35
CV3	22	27
CK	9	64
CR2	16	62
CH	4	7
TOTAL	3478	11361

[a] Table 8.1 contains a summary of the numbers of chondrites according to tabulations in Grady, M. M. (2000) *Catalogue of Meteorites* (5th edn) Cambridge: Cambridge University Press. The Catalogue lists all validated meteorites reported through December, 1999, including 17 808 antarctic specimens. Antarctic meteorites have been collected principally by Japanese, US and European field teams. The Japanese collection listed in this source contains 8501 antarctic meteorites. Of these, 2850 have been classified. 2647 of the 2850 are chondrites. The listed ANSMET collection includes 8867 specimens, of which 8850 have been classified; 8520 of the 8850 are chondrites. The listed EUROMET (European consortium) collection includes 431 meteorites, of which 397 are chondrites. The diligent reader, adding up the numbers of antarctic chondrites in this paragraph, will get a total of 11564. The difference between this number and the total in Table 8.1 represents 203 chondrites, mainly carbonaceous, whose group designation has not yet been determined, so they could not be included in this table.

begin where the chondrites leave off. Typically, they do not contain chondrules because at the temperatures necessary for complete recrystallization or melting, chondrules have been assimilated. In rare cases, a few relict chondrules are found, supporting the idea that achondrites have been formed from chondritic material. Most achondrites show evidence of igneous fractionation of the kind that would occur in a large body that had differentiated into a metallic core and silicate mantle, therefore iron meteorites and stony irons perhaps were associated with some achondrites in the same parent bodies. Isotopic age determinations on the whole rock, as well as on individual minerals within the rock, indicate that the minerals of achondrites all formed at the same time as the original core-mantle fractionation of their parent bodies (around −4.6 to −4.5 billion years). Unlike larger bodies, later igneous activity did not occur on the parent bodies of the achondrites. Their ages of formation also speak to the history of planet and parent body formation in the solar system: accumulation and fractionation all occurred at around the same early date, under extraordinary heating conditions that did not recur.

The seven currently recognized achondrite groups are as follows.

Brachinites
The group is named after a stone found in 1974 near Brachina, South Australia. Some brachinites may be residues of CI-like chondrites that underwent slight partial melting with subsequent removal of the molten phase; others may have been completely recrystallized at temperatures just below those at which melting would begin to occur. In either case, their parent body heated up, to temperatures high enough to destroy the original chondritic textures but not to temperatures high enough to cause appreciable composition change through removal of a liquid phase.

Acapulcoites/Lodranites
Acapulcoites are named after a stone that fell in 1976 near Acapulco, Mexico and lodranites are named after Lodran, a stone that fell in 1868

near Lodhran, Pakistan. These were thought earlier to be separate groups, but two antarctic meteorites are transitional between them and this seems to tie them to the same parent body. Acapulcoites are believed to have recrystallized from chondrite-like parent material with little, if any, loss of a liquid phase. Lodranites have undergone more extensive melting, with some fractionation due to migration of a basaltic liquid phase and slight loss of that mixture of iron and nickel (Fe,Ni) and iron sulphide (FeS) that melts at the lowest temperatures. The recrystallization and partial melting, just described, occurred very soon after the formation of the solar system. While three acapulcoites actually have rare chondrule relicts, indicating that they formed from chondritic material, this inferred parent material cannot be identified with any of the known chondrite classes. The relict chondrules, as well as discernible effects of partial melting without removal of the melt phase, give reasons to believe that the acapulcoite/lodranite group represents a very early stage in parent body evolution from chondrite material to achondrite material. This record is not preserved in other groups.

Ureilites

This group is named after a stone that fell in 1886 onto a farm called Novo-Urei at Karamzinka, in the former USSR. Many ureilites have been identified and described. All have the same distinctive oxygen isotope ratios and are quite high in carbon content. Because we have so many specimens of them, one would think that this group would be very well understood, but this is not the case – ureilites pose more problems of interpretation than they answer. They resemble the types of rocks that have formed by crystal accumulation on the floors of magma chambers or that have formed as residues of partial melting that was extensive enough to fractionate away liberal proportions of the original components. Such rocks are described as *ultrabasic*. Wherever we find ultrabasic rocks in nature, we expect to find associated complementary basaltic rocks. These have compositions derived from the parent magma, either by the settling out of early-formed crystals or by squeezing out of the early-formed liquids during partial melting. Both processes will fractionate the original starting material, and the result is a body of ultrabasic rock associated with a body of basaltic rock that has been derived from it.

The mystery is not why we have ultrabasic rocks like the ureilites – the mystery is why we do not find meteorites from this parent body (or bodies) with compositions representing the basaltic rocks that must have fractionated from their common parent material. Because the igneous partitioning that parent magmas undergo results in basaltic rocks near, or on, the surface and ultrabasic rocks at depth, some workers have suggested that the original basaltic crust of the ureilite parent body was eroded away by collisions with other large bodies, or that the crust was blown off the body by explosive volcanism. The problem here is that either of these processes would supply basaltic fragments, some of which could meet the earth as meteorites. So the question still is, "Where in the solar system are the ureilite-associated basalts hiding?" One possible answer is that separation of the basaltic overlayer occurred so early in solar system history that fragments blasted off the surface were collected into other bodies quite early and none of them exist in space today. If this is not the case, we will have to reconsider our model of the primitive solar nebula: maybe the ureilite parent body or bodies formed in a region of the nebula that had been somehow depleted in certain components, primarily Al_2O_3 (aluminum oxide) and SiO_2 (silica). These would have gone preferentially into the basaltic complement of the ultrabasic/basaltic pair. It is not clear, however, what type of nebular fractionation process this could have been.

Ureilites are similar to carbonaceous chondrites in their high content of carbon, so maybe they are partially melted carbonaceous chondrites. They are also similar to those carbonaceous chondrites of primitive composition (aqueous alteration stages 1 and 2) in their oxygen isotope ratios. This might indicate something about their origins, however, it would not explain the problem of the apparently missing complementary basalts.

In addition to these problems, trace element concentrations in ureilites differ enough from specimen to specimen that some workers believe the group derives from more than one parent body. If that were true, it would make an origin involving igneous fractionation less believable, because in that case, *all* the ureilite parent bodies would have had to have lost their basaltic surface materials.

Aubrites

This group is named after Aubres, a stone that fell in 1836 near Nyons, France. Aubrites are igneous rocks whose large crystal sizes suggest slow cooling in a large magma body. The minerals that form aubrites have a remarkable deficit in FeO. Some of the original FeO may have been removed in liquids produced during fractional melting, but this process alone would not cause the almost complete removal of FeO that we observe. Any remaining original FeO could have been incorporated into sulfides or reduced to metallic iron, most of which then fractionated out. There is reason to suspect that aubrites formed at unusually high temperatures, both because of the extreme chemical reduction of the minerals and also because of the presence of a highly refractory mineral, oldhamite (CaS). Enstatite $(MgSiO_3)$ is the major mineral of aubrites, and because this is also the major mineral in enstatite chondrites (EHs and ELs) it would be natural to suspect that aubrites are a residue derived by partial melting of one of those chondrite groups, followed by separation of the liquid phase. If this were the case, however, the dilemma is similar to the one involving ureilites – we have not yet found igneous meteorites that formed by crystallization of such a liquid fraction. Because of compositional similarities between enstatite chondrites and the very enstatite-rich aubrites, it is tempting to propose that they are related, but any such relation may be limited to having formed in the same general region of the primitive solar nebula. If this were true, it follows that the solar nebula was, indeed, chemically fractionated before accumulation of the meteorite parent bodies.

Howardites/eucrites/diogenites

These three meteorite groups, called HED meteorites, are related – they embrace a clear history of previous association in the same parent body. Diogenites were originally named shalkites, after the then only known member, Shalka. In 1872, Gustav Tschermak changed the name to diogenites, to honor Diogenes of Apollonia, who Tschermak thought had been the first to recognize meteorites as nonterrestrial bodies. Diogenites are probably cumulates that settled to the bottom of a magma chamber during a fractional crystallization process. Eucrites were named by Gustav Rose in 1863. The name is taken from a Greek word meaning "clearly determinable." Some eucrites are residual in appearance (i.e., they are what is left after early-formed liquid has been removed), some appear to result from accumulation of early-formed crystals, and thus are cumulates, and others appear to have crystallized from a basaltic liquid. Howardites, named after Edward Howard, the British chemist who did important early analyses of meteoritic stones, are mechanical mixtures of diogenite fragments and eucrite fragments in various proportions, and it is the howardites that suggest so strongly that these three achondrite groups derive from the same parent body. Their reflection spectra match quite well to the reflection spectrum of Asteroid 4 Vesta, located in the asteroid belt, and also to several earth-approaching asteroids that well may have been derived from Vesta.

The age determined for core-mantle separation of the HED parent body is around 4.55 billion years, which must be very close to the time when the planet itself formed, but the ages of crystallization of some of the eucrites are around 4.40 billion years, suggesting that this parent body was large enough to maintain magma chambers for 150 million years after the planet formed. All else being equal, this suggests the HED parent body was smaller than earth's moon, but larger than the parent bodies of some of the other meteorite types.

Angrites

This group is named after the type specimen Angra dos Reis, which fell in Brazil in January, 1869. They are crustal igneous rocks of basaltic composition that may have been derived by complete or partial melting of CM or CV carbonaceous chondrites. An unusual aspect of their composition is that they are greatly depleted in moderately volatile elements but less so in highly volatile ones. It is not at all obvious how this could have occurred during processing on the parent body, because the more-volatile elements should have been lost preferentially; this by definition of the term "volatile." It may be that the original igneous processing of these achondrites had depleted them in both the highly volatile, as well as the medium volatiles, and that an infusion of highly volatile elements occurred during the final stages of cooling. The existing small suite of specimens may not be representative of the group.

Winonaites

This is a small group of achondrites named after a meteorite found in an archeological site near Winona, Arizona. Winonaite inclusions are found commonly in at least one group of iron meteorites, and it is a very strong presumption that these two groups derive from the same parent body. This allows us to conclude that at least one meteorite parent body was highly differentiated into a rocky mantle and a metallic core and, if this occurred in one parent body, why not others? This stony meteorite–iron meteorite relationship also provides a smooth segue into the next sections, dealing with stony-irons and irons.

THE ROLE OF ANTARCTICA AS A SOURCE OF ACHONDRITES

With the possible exception of the HED meteorites, the number of achondrites available for study is quite limited, and for most of the achondrite classes we are not yet sure that we have even a representative collection (Table 8.2). In most cases, the numbers in the "antarctic" column exceed those in the "non-antarctic" column. This

Table 8.2. *A compilation of non-antarctic achondrites compared to the numbers of antarctic achondrite specimens that have been described*

Group	Non-antarctic	Antarctic
Brachinite	6	1
ACAP/LOD	4	22
Ureilite	29	63
Aubrite	10	34
H/E/D	85	302
Angrite	1	3
Winonaite	5	6
TOTAL	140	431

This tabulation derives from the same source as listed for Table 8.1.

is a measure of the value of our continually growing collections of antarctic meteorites.

STONY-IRON METEORITES

Since stony-iron meteorites consist of variable but significant parts of nickel-iron and stone, they must be samples from sites in differentiated parent bodies where mobile nickel-iron and crystalline components of igneous rocks were more or less equally available. But the two principal types of stony irons are quite different from each other, as follows.

Pallasites

This is a group of meteorites that typically contain olivine crystals embedded in a matrix of nickel-iron. Pallasites are generally believed to be samples from the core/mantle boundaries of their parent bodies. We know that olivine is often the first major mineral to settle to the floor of a magma chamber. It seems likely that it would be the first mineral to accumulate at the bottom of a rocky mantle, if the mantle were hot enough to be extensively melted and just beginning to cool down enough for crystallization to begin. Now in a differentiated parent body, the bottom of the mantle is the top of the metallic core. If the upper part of the core were molten also, and if the core were rotating slightly faster than the mantle, or if there were circulating convection cells in the core, there could be a turbulent zone at the core/mantle interface within which metal could mix with the overlying accumulation of olivine crystals. So pallasites may be giving us edge-on views of the core/mantle interfaces on their parent bodies.

Mesosiderites

We tend to assume that if a parent body were large enough to become even partly molten, then a metallic core would have formed, mantled by igneous rock. Mesosiderites appear to be composed of fragments of broken igneous rock which were mixed with liquid nickel-iron alloy, which then cooled and solidified before core formation could occur. But the fact that the igneous rocks are fragments of larger igneous rock masses suggests that core/mantle fractionation had occurred, and mantle materials were later fractured and intruded by the still-liquid metallic core material. One group of mesosiderites even appears to have formed from liquid silicates, mechanically mixed with liquid nickel-iron, coexisting in an uneasy alliance. Because these two phases will unmix (like oil and water), one would expect conditions to be ideal for gravitational separation in the same way that oil will float to the top of a container of water. But obviously this did not happen before solidification made it impossible.

This intriguing class of stony-iron meteorites presents other apparent contradictions. The rocky component is olivine-poor, like most achondrites. One could assume that during cooling of a largely molten mantle, olivine, the first-formed crystalline species, had sunk to the bottom. But then, when remixing of core and mantle occurred, the same accumulation of olivine would have been carried along with the molten core material and become part of the melange. To help explain the missing olivine, suggestions have been made that mesosiderites formed from primordial material that was naturally olivine-poor. This is equivalent to saying that the bulk composition from which the mesosiderites formed was richer in SiO_2 than most chondrite material, and because of this, very little olivine would crystallize out of the melt. This could require formation from material in a somewhat more SiO_2-rich part of a fractionated solar nebula.

The metal phase in mesosiderites presents another contradiction. Studies of the metallic fraction of mesosiderites indicate that the metal cooled very rapidly at first, and then extremely slowly later. This has been taken to suggest that the initial rapid cooling and solidification occurred near the surface of the parent body while the later, slow cooling occurred at depth. How this turnover of mantle material could have occurred is not known. Large impacts have been suggested, as well as sudden mantling of the surface materials by thick layers of dust, before cooling was complete. This would provide an insulating blanket and greatly slow the cooling rate.

Debates over the origin of the mesosiderite parent body, or bodies, typically invoke many more special circumstances than are necessary to visualize the conditions of origin of most other meteorite parent bodies. This simply indicates that we do not understand the contradictory evidence presented by the mesosiderites. It also allows great latitude in picking and choosing between alternate proposed histories, or taking aspects of different ones and building them into some different possible history. Using selected observations and interpretations, one could postulate a reasonable beginning to the history of these intriguing objects, and a reasonable end for that history, but the part in the middle is weak. Such a history could be a three-stage one, as follows.

- The beginning.
  If we can swallow the assumption of a SiO_2-rich source for the mesosiderites (and, thus, a somewhat fractionated solar nebula), we can assume that parent body accumulation occurred from local materials, the mass heated, core/mantle fractionation

occurred in the usual way, and not much olivine crystallized in the igneous mantle rocks.

- The middle.

  At a time when the metallic core and part of the lower mantle were still molten, *something happened that would have been fascinating to watch, from a safe distance, had we been there . . .*

- The end.

  After that happened, the mesosiderite parent body had undergone remixing between core and mantle materials and had been divided into small planetesimals, each of which began cooling rapidly under gravity-free conditions so that the metal and silicate fractions did not separate. Cooling continued until well below the temperature at which no more liquid remained. During this time the planetesimals slowly reaccreted – very gently, so that the gravitational heat of accumulation did not produce enough heat to reinitiate planet-wide melting and differentiation. Final cooling occurred at a very slow rate after reaccretion. The mesosiderite parent body survived long enough to experience the effects of whatever makes little ones out of big ones in the asteroid belts, and which sends some of the little ones to us, gratuitously.

Returning to the middle part of this hypothetical history, one must ask what type of interaction could produce the force necessary to remix core and mantle and then fragment a body of some hundreds of kilometers in diameter, without splashing the liquid fractions away as small drops and without sending the solid fragments tumbling away in different orbits, never to be reaccumulated as the mesosiderite parent body? Could a low-velocity impact with another body of equal size be that gentle? Could a close approach to the Roche limit of Jupiter accomplish these effects? The history we can read in the mesosiderites does not give us answers.

After I had written the paragraphs above, an article by E.R.D. Scott, H. Haack and S.G. Love appeared in the journal

Meteoritics and Planetary Science, suggesting a rather similar history for the formation of mesosiderites (see 'Suggested Reading'). Obviously, I think Ed Scott and his colleagues are on to something here. They suggest a low energy disruption, during which molten core material is completely mixed with overlying silicate fragments, which are cooler. Contact with the cooler rock fragments chilled the metal very rapidly at first, but as the parent body reagglomerated the cooling rate slowed markedly. They also suggest that the scarcity of olivine-rich rock fragments in mesosiderites is just an artifact of incomplete sampling, or of some process that excludes the expected olivine-rich mantle fragments from the final accumulation. This seems to me to be a rather artificial device but it is necessary if one assumes that all asteroidal parent bodies had original bulk compositions that would produce a significant region of olivine-rich rock at the mantle–core interface. On the other hand, to assume that the mesosiderite parent body accumulated in a more SiO_2-rich region of the solar nebula, so that not much olivine would form in the mantle, can also be accused of artificiality.

THE ROLE OF ANTARCTIC STONY-IRONS

It is interesting to note that the antarctic occurrences have added so few pallasites to the world's collections, while at the same time adding a much greater number of mesosiderites (Table 8.3). It is somewhat disappointing, however, that the new mesosiderites made available from Antarctica have not given us important new clues to their origin. Maybe we already have all the clues we will get, and we will be forced into flights of fancy, as outlined above, to try to explain their apparent contradictions.

THE IRONS

Iron meteorites are actually nickel–iron alloys containing varying proportions of nickel and small amounts of cobalt. They typically contain minor amounts of sulfides, phosphides and carbides as inclusions, and

Table 8.3. *A compilation of non-antarctic stony irons compared to the numbers of antarctic stony iron specimens that have been described*

Group	Non-antarctic	Antarctic
Pallasites	43	7
Mesosiderites	36	30
TOTAL	79	37

This tabulation derives from the same source as listed for Table 8.1.

in solution a range of those elements that tend to have chemistries similar to metallic iron. The crystallographic structure of the nickel–iron alloy is dependent partly on cooling rate and partly on the content of nickel present, and the original classification scheme for irons was a structural one. A chemical classification of irons was suggested more recently by John Wasson, at UCLA. This is the classification system in use today. It depends on ratios of elements present in trace concentrations in solution in the nickel–iron alloy.

In general, each chemical class of irons is considered to represent material from the core of a different meteorite parent body. It is noteworthy, however, that there are many iron meteorites that do not fit into the classification system. Because there are only one or two examples of each, they are not considered abundant enough to serve as the bases for new classes, and are lumped together under the term "ungrouped." But if each of these potential new groups came from a different parent body the estimated number of meteorite parent bodies in the early solar system might be increased by a factor of five or six.

The chemical classification of iron meteorites, depending as it does on the relative concentrations of trace elements in the parts-per-million and parts-per-billion ranges, is too technical for the purposes of this book, so it is not discussed here. Suffice it to say that the group names consist of a Roman numeral followed by one or more capitalized letters. Thus, group IVA would be read as 4A.

Some of the more interesting irons contain silicate inclusions that suggest similarities to known achondrite or chondrite classes. Classes IAB and IIICD irons, for example, contain silicate inclusions apparently identical to winonaite achondrites, and it is a reasonable presumption that in these three meteorite classes we have samples from the cores and at least the lower mantles of at least one, and perhaps two, parent bodies. How some of the core became remixed with some of the mantle materials is an unsolved problem, with some of the same aspects as the mesosiderite problem described above.

A similarly mysterious case involves the IIE irons, which contain silicate inclusions that are somewhat H-chondrite-like in composition but display very variable degrees of metamorphism, from recognizable chondrite fragments through partially melted chondrites, to totally melted chondrites that have lost their metal and sulfide components (in other words, have undergone some degree of chemical fractionation), and basaltic fragments that could be rocks crystallized from a differentiated melt of chondrite material, but showing different degrees of differentiation. Obviously, if all these rocky inclusions came from the overlying mantle the parent body had to have been very heterogeneous; moreover, whatever caused this mixing between core and mantle material was remarkably efficient at sampling many regions within the partially evolved rocky materials.

Another iron meteorite group with silicate inclusions is IVA. The included silicates in IVA meteorites are surprisingly rich in the SiO_2 component which, superficially at least, might support the idea mentioned earlier that mesosiderites had accumulated from material in a SiO_2-rich region of the solar nebula, but this source would have to have been even richer in SiO_2 than the source of the mesosiderites.

With all this remixing between cores and mantles, one could easily get the impression that the very early solar system was a strange, and sometimes, in some places, a violent scene, with parent bodies accumulating, heating up, melting on a large scale, fractionating to

Table 8.4. *A compilation of non-antarctic irons compared to the numbers of antarctic iron specimens that have been described*

Group	Non-antarctic	Antarctic
IAB	113	18
IC	11	0
IIAB	74	29
IIC	8	0
IID	16	0
IIE	17	1
IIF	5	0
IIIAB	225	5
IIICD	40	1
IIIE	13	0
IIIF	6	0
IVA	63	1
IVB	13	0
Ungrouped	79	16
TOTAL	683	71

This tabulation derives from the same source as listed for Table 8.1.

cores and mantles, being shattered or partly shattered, and with some of the fragments being slung into highly eccentric orbits to collide, at hypervelocities, with other bodies. This is probably a correct impression. It could be described as a very vigorous environment.

THE ROLE OF ANTARCTIC IRONS

Iron meteorites are much less abundant in the antarctic collections than in the non-antarctic ones, as seen in Table 8.4. This result is even more striking when it is realized that 25 of the IIAB antarctic irons are from one fall, found at Derrick Peak, and by analogy with the non-antarctic meteorites, should only be counted as one specimen.

Another unusual feature of the antarctic irons is that they contain a higher proportion of anomalous or ungrouped individuals than the non-antarctic irons. Ungrouped specimens represent 11.5% of non-antarctic irons; counting Derrick Peak as one fall, fully 34% of antarctic irons are ungrouped. Since each anomalous or ungrouped iron may represent a parent body that is not represented by other specimens in the collection, the antarctic irons may have contributed disproportionately to the total number of meteorite parent bodies that could be estimated.

SUMMARIZING THE ROLE OF ANTARCTIC ASTEROIDAL METEORITES

In many cases, the antarctic meteorite collection has simply added numbers of new specimens to a class of meteorite that was already abundant in our collections. In other cases, such as those of the rarer R, K, CI, CK, CR2 and CH chondrites, and the acapulcoite/lodranite, angrite and winonaite achondrites, the addition of new specimens to the existing collections has been significant. In the cases of the pallasites and all of the irons, the new specimens found in the antarctic collections have been many fewer than might have been expected, based on the numbers already in our collections. An interesting contradiction with regard to the irons, however, is that while we have found fewer irons in Antarctica than we might have expected, a very large proportion of those few are ungrouped irons, furnishing possible clues to the existence of a much larger number of parent bodies than we might otherwise have postulated.

THE ASTEROID BELT

We have seen that chondrites, as undifferentiated meteorites, can furnish clues to conditions within the primordial cloud, to conditions within the solar nebula that formed from it, and to the history of small bodies that formed within the solar nebula. Irons, stony irons and achondrites, as differentiated meteorites, can give us hints about the deeper, hotter regions of parent bodies that have generated enough heat to differentiate into core-mantle associations shortly after their formation. They can be contrasted to lunar and martian meteorites, which can attest to conditions on bodies that were large enough to continue to generate igneous activity for long periods after they were formed. The large numbers of meteorites now being recovered in Antarctica add force to our inferences about the asteroid belt.

One goal is to use meteorites of asteroidal origin to try to construct internally consistent parent-body models and, eventually, try to estimate the number of meteorite parent bodies that originally existed.

Another goal that we can now begin to think about is to understand that region of the solar nebula within which the asteroidal meteorites formed and follow its history until the present. Some of the clues we can glean from the study of the meteorites are presented:

- Meteorite parent bodies were not chemically identical, so chemical gradients may have existed across the asteroid belt.
- The existence of iron, stony and stony-iron meteorites indicates that at least some parent bodies were large enough to generate enough heat during and shortly after formation to be able to differentiate into a metallic core and a rocky mantle. The larger asteroid parent bodies were hot enough to generate igneous rocks at depth and cause varying degrees of thermal metamorphism at lesser depths.
- We do not find igneous meteorites younger than about 4.3 billion years, so no asteroid parent bodies grew large enough to prolong igneous activity much beyond their period of formation. Thus, none grew as large as Mars, for example, unless those larger bodies were destroyed very early on, before they could generate younger magmas.
- We find many, many "anomalous" irons. This suggests that there were many, many meteorite parent bodies in the asteroid belt, each large enough to have undergone core/mantle differentiation.

- Many achondrite classes in our collections represent rock types that would have been produced as *residues* of igneous differentiation, but our collections lack corresponding specimens that would have resulted as *fractionated products* of the differentiation process. This suggests that meteorite parent bodies often came close enough to collide. If such collisions did not result in mutual fragmentation, they probably caused crushing, removal and dissipation of the outer layers from each body. Near misses could have caused gravitational disruption, followed in some cases by reagglomeration. Other close approaches could have caused mutual perturbation of the former orbits of both bodies. The asteroid belt was a chaotic region at the time the solar system formed.

One may, or may not, accept this picture of the early days of the asteroid belt, derived from study of the meteorites we have found. It would be nice if there were independent support for these ideas. As it happens, there is such support.

John Chambers at the Armagh Observatory in Ireland and NASA Ames Research Center, Moffett Field, California and George Wetherill at the Carnegie Institution of Washington have proposed a model history of the asteroid belt, based entirely on physical principles and extrapolating backward in time from the present.

Before I go into a short description of their model, the general reader should understand that physicists like to start with the fundamental laws of physics and derive general models that can be applied to the real world. Even though they are uniformly very smart, they sometimes make wrong conclusions based on their models. Following are a couple of examples.

In Chapter 6, I explained that many meteoriticists devoutly wished to believe that the SNC meteorites originated on Mars, in spite of the claims of physicists that the energy density sufficient to levitate a 10 kg rock off the martian surface and into orbit about the sun would shock it to such a degree that it would either be pulverized, or melted and dispersed as droplets. In Chapter 7, I outlined our triumphant glee at finding the first lunar meteorite, and noting that it was almost completely unshocked, even though it had been ejected from the lunar surface by an impact.

"Aha!" I thought. "Now we have them where we want them!" (Whatever that means...) In short order, however, one of them returned to his impact model and discovered a region outside the impact site where shock waves underwent destructive interference and from which rocks could be propelled off the surface and into orbit, and show no shock effects at all! It was there in his model, all the time, but before he realized it, many of us had lost a lot of hair.

On another occasion, we noted that some of the meteorites we were finding on the antarctic ice sheet had fallen as much as 1–2 million years ago. Some of us came to believe that, by comparing the very old falls with the more recent ones, we might find evidence that the character of the meteorite flux reaching the earth had changed over time. We discussed having a workshop devoted to this question, in spite of the fact that one of the authors cited above (G.W.) had already shown that the mechanisms that eject fragments from the asteroid belt could cause orbits to evolve only over tens or hundreds of millions of years. We thought that it would be nice to have this viewpoint represented at the workshop, so I called George and asked him to be on the program. He said that he didn't know what he would say that was new, since he had already said it all in his published paper. I suggested that he could give a talk in which he discussed which of his basic assumptions could be wrong, in case we found that the flux *had* changed with time. He pretended to give this some consideration, and then said,

"Well, that would have to be F = ma."

After that, of course, there was nothing more to be said. But there are some anomalies that appear when one compares the relative abundances of antarctic meteorites and modern falls. The more obvious of these are among the iron meteorites and, as I show in the next chapter, among the HED achondrites. It could be that changes in the flux of meteorites reaching the earth *can* occur over periods of much less than a million years, and it might be wrong to assume that all meteorites reach the earth by the same route.

This is not to say that physicists are always wrong and geologists are always right. The natural philosopher prefers to reach general conclusions based on specific observations, and is often wrong. The best situation is one in which general conclusions based upon observations converge toward agreement with general conclusions based upon the fundamental laws of physics including, of course,

Force = mass × acceleration.

I was pleasantly surprised and greatly encouraged, therefore, when I read the paper by Chambers and Wetherill (see 'Suggested Reading') and realized that all the clues the meteorites had been giving us fit very nicely, or at least were not in conflict with, the picture they painted based on the fundamental laws of physics.

Beginning early in the existence of the solar system, they suggest an asteroid belt teeming with small bodies, most of which have been removed over the last 4.2–4.3 billion years, or so, either by falling into the sun or being slung out of the solar system. The bad boys in this scenario are the giant planets, Jupiter and Saturn. If an asteroid, which has almost no mass in comparison to these planets, gets a repeated gravitational tug at regular intervals, always at the same point in its orbit, the cumulative effect can be to speed it up or slow it down, depending upon whether the tug is from ahead or from behind. If the tugs tend to speed it up, part of the asteroid's orbit extends farther and farther away from the sun, and eventually it becomes Jupiter-crossing, then Saturn-crossing. It may crash onto one of these planets, or it may be thrown out of the solar system, in the same way that we can use a planet to speed up an artificial satellite (the "slingshot effect"). On the other hand, if repeated gravitational tugs from the major planets tend to slow it down, one part of the asteroid's orbit remains in the asteroid belt, but that part of the orbit that passes closest to the sun migrates closer and closer to the sun, becoming Mars-crossing, then Earth-crossing, Venus-crossing, and so on. Eventually it crashes into one of the inner planets, or into the sun itself. In both cases, speeding up or slowing, it is removed from the asteroid belt over a long period of time.

In order to receive these tugs, always in the same direction and always at the same point in its orbit, an asteroid must complete its orbit during a period (a *resonance frequency*) that is some simple fraction of the period of Jupiter's orbit. To have such an orbital period, an asteroid's orbit must have one of several radii. If it is orbiting the sun at one of these radii it will certainly be removed from the asteroid belt. It's just a matter of time. If, in addition, the asteroid belt is densely populated with asteroids that are perturbing each other's orbits by close approaches and, by collisions, creating more and more fragments, each in its own new orbit, material can be supplied to the resonance frequencies and ejected from the asteroid belt with greater efficiency.

In a collision between a golf ball and a bowling ball, the bowling ball will win. But if the same bowling ball collides with many golf balls, all in the same direction, it may notice that something is happening. Jupiter and Saturn, of course, give up a little energy any time they perturb an asteroidal fragment, and their orbits have become more circular as a result.

Chambers and Wetherill have run computer simulations in which they assumed the presence of large numbers of asteroids of a variety of sizes on nearly circular, nearly coplanar orbits spanning the asteroid belt. They have shown, to the complete satisfaction of their participating computers, that there would be only two or three major asteroids remaining after some 200–800 million years, depending on the starting conditions that were assumed. Because of frequent disruptive collisions, and masses adopting resonance frequencies and then being ejected from the asteroid belt, no single, large planet could successfully accumulate.

From the clues we read in the asteroidal meteorites and the calculations of Chambers and Wetherill, as well as the speculations of many individuals over recent decades, we can formulate the following picture of events in the region between Mars and Jupiter.

When the rarefied primordial cloud collapsed and became the somewhat denser solar nebula, planet building began. How to accomplish the first stage of this process – getting bodies to grow large enough to become self-gravitating – is still a source of speculation. Initial aggregations may have been held together by surface forces between tiny particles that had condensed from vapor, but such agglomerations might not be able to grow beyond a small size, particularly if the mass were spinning. In 1963, Bert Donn, at the Goddard Space Flight Center, suggested that crystallites condensed from vapor would be extremely irregular in shape. Many would be quite whiskery and could interlock with each other, so that quite a large body might be physically strong enough to hold together – like a bucketful of fishhooks (analogy mine). By whatever means the initial accumulations of grains were held together, fluffy agglomerations of grains formed large bodies.

When these low-density accumulations became large enough, they started adding to their mass by mutual gravitational attraction between them and other bodies in their vicinities. Growth also occurred by gentle collisions or, more precisely, collisional interpenetration between large, low-density bodies. As they gained more mass, they became more compact and as a function of their increasing masses, were able to attract additional material from farther away.

This was a relatively quiet time, but high energy reactions were imminent and violence loomed on the horizon. With continuing growth, densities of the forming planets increased and radioactive heating began to be effective. This was supplemented by collisional heating from increasingly violent encounters with other dense bodies. Once temperatures became high enough to melt iron, heat was liberated by the conversion of potential energy to kinetic energy by masses of molten iron falling toward the center of the body. In the asteroid belt, the next stage should have been growth of one or two large planets from many small ones by gentle collisions or, more probably, by gravitational aggregation of the fragments produced by collisions, either onto a surviving body or toward the center of mass of the cloud of fragments.

That is what should have happened, except that, off to one side, Jupiter and Saturn had been growing and were now bulking large upon the scene. As a result, the forces of planet growth in the asteroid belt, working upon the gas, dust, growing planetoids and collision fragments, went to war with the forces of depletion, represented by gravitational tugs that very efficiently removed any matter in the resonance frequencies. The giant planets won. Relatively speaking, there is very little left today in the asteroid belt, but those bits and pieces that are still being removed and that happen to encounter the surface of the earth are refugees from the violence which give mute testimony to the nature of the conflict. If you can visualize the entire battle speeded up; taking place almost in the time necessary to relate this history, you might try imagining it while listening to Tchaikovsky's 1812 Overture. Follow the frenzy of battle, the shifting tides of relative advantage, the swirling eddies of intense activity and the calmness of anticlimax. It works for me . . .

SUGGESTED READING

Brearley, A.J. and Jones, R.H. (1998) Chondritic meteorites, Ch 3. In *Planetary Materials*, ed. J.J. Papike, *Reviews in Mineralogy* vol. 36. Washington, D.C.: Mineralogical Society of America.

Cameron, A.G.W. (2001) From interstellar gas to the earth-moon system. *Meteoritics and Planetary Science* **36**, 9–22.

Chambers, J.E. and Wetherill, G.W. (2001) Planets in the asteroid belt. *Meteoritics and Planetary Science* **36**, 381–399.

Mittlefehldt, D.W., McCoy, T.J. Goodrich, C.A. and Kracher, A. (1998) Nonchondritic meteorites from asteroidal bodies, Ch 4. In *Planetary Materials*, ed. J.J. Papike, *Reviews in Mineralogy* vol. 36. Washington, D.C.: Mineralogical Society of America.

McSween, H.Y., Jr. (1999) *Meteorites and Their Parent Planets*, 2nd edn. Cambridge: Cambridge University Press.

Scott, E.R.D., Haack, H. and Love, S.G. (2001) Formation of mesosiderites by fragmentation and reaccretion of a large differentiated asteroid. *Meteoritics and Planetary Science* **36**, 869–881.

Excellent color photos of antarctic meteorites can be found in the following publications:

Yanai, K. (1981) *Photographic Catalog of the Selected Antarctic Meteorites*, Tokyo: National Institute of Polar Research.

Yanai, K. and Hideyasu, K. (1987) *Photographic Catalog of the Antarctic Meteorites*, Tokyo: National Institute of Polar Research.

For those wishing to learn more about the fascinating history of discovery involving dust grains that form in the atmospheres of expanding red giant stars and the expanding envelopes of matter around supernovas, the following reference is a good place to start:

Bernatowicz, T.J. and Walker R.M. (1997) Ancient Stardust in the Laboratory. *Physics Today* **50**, 26–32.

Part III   Has it been worthwhile?

After 31 years during which the Japanese antarctic meteorite program has existed, 25 years of the ANSMET program, and three or four field seasons by European consortia, we have a collection of around 30 000 meteorite specimens from Antarctica. Is there anything we can learn from this massive group of specimens that we could not have learned by patiently waiting for falls to occur during the normal course of events, in more convenient regions of the world? We have seen already that one martian meteorite from Antarctica has had an important role in establishing Mars as the planet of origin of the SNC meteorites. Our first lunar meteorite, also found in Antarctica, has provided proof that rocks can be lofted into space, at least from smaller planetary bodies. The antarctic collection has added significantly to the numbers of available specimens of some of the rarer meteorite groups. But all this would have occurred sooner or later over the careers of successive generations of planetologists, patiently waiting for new falls to arrive. So the question is, "Is there anything unique about the antarctic collection to justify all the effort that has been expended in amassing it?"

I have given a lot of thought to this question, and I believe the answer is yes, for the following reasons.

- There is much to be said in favor of systematic collection and controlled curation of large numbers of meteorites in a frozen environment such as that of Antarctica. We can get an independent estimate of the proportions of the different groups of meteorites reaching Earth.
- It is useful to know that we now have samples of stony meteorites that fell as long ago as two million years. Stony

meteorites are not preserved for nearly that long outside Antarctica, with the possible exception of Greenland.

- During their residence times on the earth, antarctic meteorites have interacted with the ice sheet, either by direct fall onto a stranding surface or by being transported from their place of fall to a stranding surface. They may be telling us something about the history of the ice sheet.

- Antarctic meteorites have been a stimulus to creative thought. There is new interest in the role of the ice sheet as a collector of micrometeorites and cosmic dust particles. Roboticists have toyed with the idea of robotic meteorite collection in Antarctica, and have taken some tentative steps in that direction.

I discuss these topics in the final section of this book, and suggest also a few other places where we might well find unique collections of meteorites and cosmic dust, but have not yet looked.

9 Evaluating the collection – and speculating on its significance

INTRODUCTION

Asteroidal meteorites form the bulk of the collection. The fact that we can find lunar and martian samples in Antarctica has been a very nice dividend for the ANSMET project and has helped significantly in ensuring a continuation of its funding over many years. But these samples are isolated faces in an enormous crowd – memorable and important, true, but very few in number. Almost all antarctic meteorites (and also meteorites fallen in the rest of the world) are believed to be asteroidal meteorites. By this we mean that they are fragments of larger bodies whose abode is (or was) the asteroid belt. The asteroid belt is a region between the orbits of Mars and Jupiter within which we have telescopic evidence of the existence of thousands of bodies in orbit about the sun. All the bodies we have detected telescopically, of course, are larger than the bodies we have collected on the earth as meteorites. If there are thousands of asteroids large enough to see from Earth with a telescope, there must be millions or billions of meteoroid-size particles there, too small to be seen, but each following its individual path in orbit about the sun. What we have in our meteorite collections is a tiny sample of all the meteorites whose orbits have, for one reason or another, become earth-crossing. These earth-crossers, in turn, are only a small fraction of the numbers of meteoroids that must remain in the asteroid belt. There is a lot of room in the asteroid belt. It spans about twice the distance between the earth and the sun, a distance that we call one astronomical unit (AU): $1\,\mathrm{AU} \approx 1\,500\,000\,\mathrm{km}$, so the asteroid belt is about $3\,000\,000\,\mathrm{km}$ across.

Chapter 8 ended with a description of the chaotic processes that resulted in the asteroid belt of today. In collecting asteroid fragments that have reached the earth, we have an intellectual challenge, and an

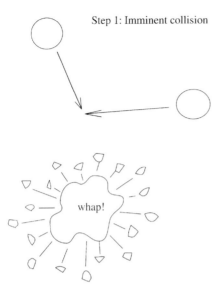

Step 1: Imminent collision

whap!

Step 2: Collision & mutual destruction

FIGURE 9.1 Each fragment records a stage in the evolution of its parent body up until the moment of impact.

opportunity, to examine bits and pieces of bodies as they existed and evolved during the planet-forming process and were then destroyed. Just as a clock on the dashboard of an automobile may stop at the instant of a collision, giving us the exact time at which the accident occurred, meteorites are clocks in the planet-forming process. Some were stopped early in the process as their parent bodies stopped growing or suddenly lacked any further practical significance by reason of a disruptive collision with another body. Others record a longer history involving accumulation, heating up and possible chemical fractionation – and then the almost inevitable, clock-stopping impact (see Figure 9.1). The parent bodies of some meteorites may have died a slower death by gradual attrition from the surface inward in a series of impacts from smaller bodies. In Antarctica we recover the stopped clocks.

The concept of Antarctica as a previously untapped source of meteorites has evolved only over the past several decades. The realization that, if only one can overcome the problems of extreme cold and isolation, one can collect hundreds, even thousands, of new meteorite specimens, has been mind-boggling. That the antarctic collection so far has added new specimens to some of the rarest groups of meteorites is undeniable. That it will continue to make this contribution is a valid assumption. So much for the facts upon which everyone agrees – I like to speculate about the messages the antarctic meteorite collection *as a whole* is sending us. This chapter is a summary of my thinking about these messages.

Antarctic meteorites have been my preoccupation for years. My thoughts about their significance have evolved over time. Some ideas may be useful and some may be proven wrong. In this chapter, I describe a personal journey through a series of those ideas, in the conceit that it will prove as interesting to the reader as it has been to me. So I issue an invitation to the reader to join me in exploring the potential value of antarctic meteorites in addressing important scientific questions. We will deal with the types and abundances of antarctic meteorites and compare them to the collection of meteorites that have been observed to fall over the past few years (the modern falls collection). We will then try our hands at developing interpretations for what we see in the antarctic collection and point out the areas that need additional information or observations. While I favor my own explanations, the reader should feel free to form his/her own opinions. Among the questions about which we will speculate are the following:

- Can we compare the antarctic finds with the modern falls in any meaningful way? (This is more difficult than it may seem at first.)
- Has the flux of meteorites in the earth's vicinity changed recently? (Even to ask this question flies in the face of current dogma.)
- Can the antarctic meteorites tell us anything about the asteroid belt and its history that we would not have known without them? (Maybe we can find some new insights.)

The basic data with which we will work are included as Tables 9.1 and 9.2. These data will be referred to in later sections, and tables derived from them will be generated.

Table 9.1. *Numbers of modern falls compared to numbers (N) and masses of ANSMET finds*

Types and groups	Modern falls N	ANSMET finds[a] N	Mass (g)[b]
Stones			
EH	8	53	1730.0
EL	7	21	3104.4
H	314	3545	608113.4
L[c]	342	3914	1156542.4
LL	72	646	41216.9
R	1	7	625.1
K	1	1	23.1
CI	5	0	0.0
CM2	15	70	6947.5
CO3	5	9	2566.2
CV3	6	20	1114.1
CK	2	61	3056.6
CR2	3	55	1537.8
CH	0	7	99.9
BRACH	0	1	4.6
ACAP/LOD	2	12	1150.5
URE	5	44	10907.5
AUB	9	35	4009.8
HOW/EUC/DIO	55	139	34289.6
ANGR	1	2	7.5
WIN	1	1	11.3
UNGR[d]	41	79	624.1
Subtotals	895	8722	1877682.3
Irons[b]			
IAB	5	15	49951.8
IC	0	0	0.0
IIAB	6	3	17873.9
IIC	0	0	0.0

Table 9.1. (cont.)

Types	Modern falls	ANSMET finds[a]	
and groups	N	N	Mass (g)[b]
IID	3	0	0.0
IIE	1	1	15.2
IIF	1	0	0.0
IIIAB	11	4	7709.3
IIICD	3	1	21.1
IIIE	0	0	0.0
IIIF	0	0	0.0
IVA	4	1	2789.0
IVB	0	0	0.0
UNGR	6	14	5213.0
Subtotals	40	39	83573.3
Stony irons			
MES	7	27	31528.5
PAL	5	4	53.5
Subtotals	12	31	31582.0
Totals	947	8792	1992837.6

[a] I give only ANSMET data. Not used are data from the EUROMET and Italian collections at Frontier Mountains. In these collections it is not clear that all specimens have been described. The huge Japanese collection is omitted because it is known that a large fraction of the specimens have not been described, and inclusion of the ones that have been described would create a biased sample. The *Catalogue of Meteorites* (5th edn) indicates that as of December 31, 1999 there were 17 808 antarctic meteorites. Of this number, 8792 ANSMET meteorites weighing 1993 kg have been characterized. The great majority of the rest are found in the Japanese collection, but a large proportion of these have been characterized only by weighing and have not been identified as to group.

(cont.)

Table 9.1. (*footnote cont.*)

The curation philosophy followed in the Japanese establishment is to sort through the thousands of meteorites they have collected, which are mainly ordinary chondrites, and pick out the rarer types for initial description and further research. This guarantees that the rarer, more interesting types can be reported and described promptly, but most of the ordinary chondrites are listed only as "stony meteorites." By contrast, the philosophy followed by NASA is to classify *all* the specimens collected by the ANSMET teams, including all the ordinary chondrites. While this involves a great deal more effort, it ensures that reliable numbers are available to use in comparing abundances of the different types and groups of meteorites.

[b] Remember that when *non-antarctic* meteorite finds are compared to falls, irons are much more abundant among the finds because irons last longer than stones in deserts. Since Antarctica is a desert, and all antarctic meteorites are finds, I assume a similar situation exists there. Many of the antarctic irons have been found on rocky slopes and in moraines, i.e., areas that would not be identified as stranding surfaces (and areas where we do not find stony meteorites). I infer a direct analogy to the situation with non-antarctic finds, and have excluded some of the antarctic irons from the tabulation. The ANSMET irons not tabulated in Table 9.1 are 25 Derrick Peak specimens, and the single Inland Forts iron. Derrick Peak was found on a rocky mountainside and Inland Forts in a glacial moraine. Their inclusion in the tabulation would affect both the number and mass of irons in the ANSMET collection in a significant way.

[c] Includes 34 paired L6 chondrites among the antarctic finds that were listed as a single 407 kg specimen (ALHA 76009) in the *Catalogue of Meteorites* (5th edn). Because fall-pairing is impossible in most cases involving antarctic meteorites, I list all fragments as individuals.

[d] Includes 73 C2 chondrites that are not further classified as CM2 or CR2 groups in the *Catalogue of Meteorites* (5th edn).

Grady. M. M. (2000) *Catalogue of Meteorites* (5th edn). Cambridge: Cambridge University Press.

Table 9.2. *Relative abundances derived from the data in Table 9.1*

Types	Modern falls	ANSMET finds[a]	
and groups	N (%)	N (%)	Mass (%)
Stones			
EH	0.8	0.6	0.1
EL	0.7	0.2	0.2
H	33.2	40.3	30.5
L[c]	36.1	44.5	58.0
LL	7.6	7.3	2.1
R	0.1	0.1	0.0
K	0.1	0.0	0.0
CI	0.5	0.0	0.0
CM2	1.6	0.8	0.3
CO3	0.5	0.1	0.1
CV3	0.6	0.2	0.1
CK	0.2	0.7	0.2
CR2	0.3	0.6	0.1
CH	0.0	0.1	0.0
BRACH	0.0	0.0	0.0
ACAP/LOD	0.2	0.1	0.1
URE	0.5	0.5	0.5
AUB	1.0	0.4	0.2
HOW/EUC/DIO	5.8	1.6	1.7
ANGR	0.1	0.0	0.0
WIN	0.1	0.0	0.0
UNGR[d]	4.3	0.9	0.0
Subtotals	94.5	99.2	94.2
Irons[b]			
IAB	0.5	0.2	2.5
IC	0.0	0.0	0.0
IIAB	0.6	0.0	0.9
IIC	0.0	0.0	0.0

Table 9.2. (cont.)

Types	Modern falls	ANSMET finds[a]	
and groups	N (%)	N (%)	Mass (%)
IID	0.3	0.0	0.0
IIE	0.1	0.0	0.0
IIF	0.1	0.0	0.0
IIIAB	1.2	0.0	0.4
IIICD	0.3	0.0	0.0
IIIE	0.0	0.0	0.0
IIIF	0.0	0.0	0.0
IVA	0.4	0.0	0.1
IVB	0.0	0.0	0.0
UNGR	0.6	0.2	0.3
Subtotals	4.2	0.4	4.2
Stony irons			
MES	0.7	0.3	1.6
PAL	0.5	0.0	0.0
Subtotals	1.3	0.4	1.6
Totals	100.0	100.0	100.0

Notes for Table 9.1 apply.

CAN WE COMPARE THE ANTARCTIC FINDS WITH THE MODERN FALLS IN ANY MEANINGFUL WAY?

One way we can try to do this is by comparing the abundances of each type and group between the two collections. Antarctic meteorites could give us an independent source of data to use in estimating the character of the flux of meteoroids in the earth's vicinity, and, therefore, of its source materials in the asteroid belt. But we must try to be sure we are comparing comparable things. If we do not understand the data we will not understand the conclusions, and if we make interpretations that have embedded assumptions, and if we do not realize those assumptions are there, our interpretations may suffer.

How do we estimate the relative abundances of meteorites reaching the earth?

The first attempts to estimate the flux of meteorites in the earth's vicinity relied upon tabulation of all museum specimens. This method produced an apparent overabundance of irons. It was noted that irons are much less abundant among *observed meteorite falls* than they are in museum collections. The cause of this difference was eventually found to be that weathering processes in the desert environment destroy stones much more rapidly than irons. Many of the irons in museum collections had been found in deserts, and this had biased the collection. To eliminate this kind of bias, therefore, estimates of the abundances of meteorites in the earth's vicinity are based on counting only the number of observed falls of each type.

Counting only the modern falls eliminated the bias due to relative weathering rates, because freshly fallen meteorites have not had a chance to degrade. Still, using this system, a 100 g meteorite counts the same as a 300 000 g specimen, if they are separate falls. Each counts as one fall. There is practical justification for this procedure. Many, if not most, falls are single individuals but many falls produce showers of fragments, and we can never collect them all. The Sikhote Alin fall in Russia in 1948 was an iron meteorite that fragmented in the atmosphere and produced around 100 small impact craters, as well as swarms of smaller fragments. The total weight of collected fragments was 23 000 kg, and we have no way of knowing how many individuals have never been found at the site where the meteorite fell, in the uninhabited forest in the Sikhote Alin Mountains to the west of Vladivostok. The Holbrook fall in Navajo County, Arizona in 1912 produced an estimated 14 000 individual L6 stony meteorites weighing an estimated 218 kg, total. The largest recovered specimen of Holbrook weighed 6.6 kg and the smallest ones weighed only around 0.2 g. Over 5000 individual stones were collected. These are extreme

	MODERN FALLS	ANTARCTIC FINDS
NUMBERS	KNOWN	UNKNOWN
MASSES	UNKNOWN	KNOWN

FIGURE 9.2 What is known and not known about modern falls and falls represented among antarctic finds.

examples. Most showers are not this large, and for most showers we do not recover a large enough number of fragments to be able to make believable estimates of total numbers or masses, so counting falls is about all we can do. Additionally, finding only one fragment from a shower is all that is needed to identify the type of meteorite causing that shower.

The antarctic collection is quite different, in the sense that none of the specimens is an observed fall, and we can only guess at how many falls the collection represents. But the stranding surfaces of Antarctica are not the fields and woodlands of, for example, Pennsylvania: in Antarctica, virtually all the fragments lying on a patch of ice can be collected and weighed.

So for *non-antarctic* meteorites we can estimate how many meteorites of each type fall, but we will never know the masses of each type falling. For *antarctic* meteorites we cannot know how many falls are represented but we can measure the masses that have fallen for each type. This is a Mr. & Mrs. Sprat type situation.[1] The two sources of data ought to have the potential to complement each other. The situation is shown in Figure 9.2.

The fundamental nature of mass

For a long time I have thought that determining the relative masses that have fallen, as we can do with antarctic meteorites but not with observed falls, yields data that are in some sense "more fundamental" than just numbers of falls. One could make an assumption that the

---

[1] Jack Sprat could eat no fat; his wife could eat no lean and so, between them both, they licked the platter clean.

two collections have identical abundances of the various meteorite types and groups. In that case, abundance ratios based on masses in the antarctic meteoroid collection could add significance to estimates of the meteoroid flux in the earth's neighborhood – we would have both number percentages and mass percentages as described in Figure 9.2. I took the position, however, that we had not yet proved that the modern flux is identical to the flux over the last several hundred thousand years.

I wanted a way to get from abundances by mass among the antarctic meteorites to abundances by number of falls so that I could compare abundances by number in both collections. Under the influence of some very fuzzy thinking I decided that if the collection is large enough, relative abundances by mass should equal relative abundances by number. This was wrong, for at least two reasons. One reason has to do with differences in density between the several meteorite types. Irons, for example, have about twice the density of stones, so the same number of irons and stones will give irons twice the abundance by mass as by number. Later, I will point out the second reason why abundance by mass may not equal abundance by number.

A concept that shows more promise does not depend at all upon mass. It is based on the assumption that *the relative abundances by number of meteorite types falling onto Antarctica is reflected in the proportions by number among fragments of the various types on the surface.* The only difference would be that the numbers on the surface would be larger than the number of falls, due to breakup in the atmosphere, breakup of those that struck a hard surface, and subsequent disaggregation by chemical weathering. If the proportions of meteorite types and groups determined by numbers of specimens on the ice reflects the proportions of meteorite species falling onto Antarctica, then by implication these would be the same proportions falling over the rest of the world. As a result, we should be able to mark the upper right-hand box in Figure 9.2 as "KNOWN," at least as a proxy for the purpose of making comparisons with the modern falls.

To summarize: abundances determined from modern fall numbers = abundances determined from antarctic finds numbers ≠ abundances determined from antarctic finds masses. We will see how this reasoning worked out in the following pages.

HAS THE FLUX OF METEORITES IN THE EARTH'S VICINITY CHANGED RECENTLY?

The modern falls and the antarctic collection do not reflect the same length of time over which data have been collected. Falls have been counted for about 200 years, and the antarctic ice sheet gives us meteorites that have fallen over the last one to two million years (but most have been on the earth only during the last several hundred thousand years). We know this about the collection because the terrestrial ages of meteorites can be measured. So the antarctic collection has sampled the flux of meteorites in the earth's vicinity for, say, 300 000 years – a much longer period than represented by the modern falls. If the flux of meteorites in the earth's vicinity has changed recently, then a comparison of the two collections might reflect this.

But why would we even look for rapid changes in the flux when we know that resonance – linked perturbation of asteroidal orbits can bring about change in the flux of meteorites arriving at the earth only over a period of several million years? Experience has taught me skepticism. Remember that early reviewers of proposals to search for meteorites in Antarctica predicted that large numbers of meteorites would not be found, and, later, respected physicists pointed out that rocks could not be ejected from the lunar or martian surfaces without becoming pulverized or vaporized. In both cases, the assumptions underlying their conclusions were faulty, and it could happen here, too. So we must ask ourselves: Is it safe to assume that this is the only way in which meteorites can reach us from the asteroid belt? Are there other methods by which the flux of meteorites at the earth could be changed over shorter time periods? If comparison of the two collections suggests recent changes in the flux, then we must ask ourselves how this could occur.

Table 9.3. *These data summarize the subtotals in Table 9.1, representing the meteorite types in the modern falls and in the ANSMET collection*

Types	Modern falls	ANSMET finds[a]	
	N	N	Mass (g)
Stones	895	8722	1877682.3
Irons[b]	40	39	83573.3
Stony irons	12	31	31582.0
Totals	947	8792	1992837.6

Notes for Table 9.1 apply.

Table 9.4. *These data summarize the subtotals in Table 9.2, representing the relative abundances of the three meteorite types, as calculated from numbers and masses*

Types	Modern falls	ANSMET finds[a]	
	$N\,(\%)$	$N\,(\%)$	Mass (%)
Stones	94.5	99.2	94.2
Irons[b]	4.2	0.4	4.2
Stony irons	1.3	0.4	1.6
Totals	100.0	100.0	100.0

Notes for Table 9.1 apply.

Do the antarctic meteorites suggest different relative abundances than the modern falls?

The first attempt at a comparison ought to be between the three meteorite types. These are the major divisions: stones, irons and stony irons. The data to be used in comparing meteorite types are abstracted from Tables 9.1 and 9.2 and summarized as Tables 9.3 and 9.4.

In the preceding tables I have presented data useful in comparing meteorite collections, both by number percentages of modern falls and number percentages of fragments in the ANSMET collection. The first thing we can notice is that the relative number abundances of the three types are different. Many workers incline to the view that irons and stony irons are indeed less abundant in the antarctic collection. A suggestion commonly put forward to explain this apparent difference is that perhaps irons are less buoyant on ice than stones and, because of the large difference in density between irons and frozen water, irons would sink below the surface and not be found. This would make them less abundant in the antarctic collections than they should be. Takesi Nagata, however, has calculated this effect for ice temperatures at 2000 m elevation in Antarctica, where the meteorite concentrations are found, and finds that only irons with a diameter greater than 1 m would sink in the ice of a stranding surface. Stony irons would have to be substantially larger to suffer the same fate. Compounding the dilemma is the observation outlined in Chapter 8, that even though antarctic irons appear to be less abundant than irons in modern falls, the antarctic collection contains a disproportionately larger number of rare and ungrouped irons than the modern falls collection. This has suggested to many workers that, in respect to irons and stony irons, the modern falls are different from the flux of these objects over the last 300 000 years, or longer.

Before comparing the two collections, one should emphasize that, all else being equal, two sets of numbers are comparable only if sample sizes are large enough. Another way to say this is that if you have a mixture of lima beans and kidney beans, and lima beans make up 99.99% of the total, by counting beans you could very quickly determine that lima beans compose 99–100% of the total, but you might have to sample a hill of (mainly lima) beans before finding enough kidney beans to reflect their true proportion in the mixture.

The numbers of stony-iron meteorites and iron meteorites in the ANSMET collection are small – these are the "kidney beans" in the overall collection, and we can suspect from their numbers that we have not collected a large enough sample of antarctic meteorites to give the true proportions of these two meteorite types on antarctic ice.

Even so, and in spite of their small numbers in the ANSMET collection, which limits any quantitative conclusions about their abundance, many people believe that even if their true proportions were known, the irons and stony irons would be less abundant in the antarctic collection than among the modern falls. There seems to be no completely convincing explanation for this, but we will see that we do not have enough data to support such a conclusion in a *statistical sense.*

Statistics rears its ugly head, and many readers have just thrown this book at the wall

People seem to react badly to statistics. Some report psychic pain. But think: how many of us experienced pain when I described a statistical study of a hill of mixed beans, *without using the dreaded s-word*? The only difference between that example and the antarctic meteorites is that the meteorites are spread out over the ice, instead of being heaped up in a pile.

So an important question is, "Are the masses and numbers listed in Table 9.3 for the ANSMET collection typical of the relative masses and numbers of each meteorite type lying on the ice in Antarctica?" If they are not, we must continue enlarging the collection until we have measured the true proportions of each type because if we have not, we certainly cannot say we have a true sample of the meteorites that have fallen. A statistician would say, *"All else being equal,* are the sample sizes large enough to represent the entire population of meteorites on the ice?" The term "all else being equal…" is a simple-sounding, yet overtly troublesome, phrase. I had originally formulated the question above without including this phrase and, next to "A statistician would say…" a statistician friend wrote, *"No we wouldn't."* Inclusion of "All else being equal…" made him happy. He pointed out, quite correctly, that there might be factors of which I was not aware that would affect the collection to make it non-representative of the population from which it had been drawn, and therefore biased.

I have been able to think of only two cases where *personal* bias might have been introduced. In one case, a field team member decided that meteorites below a certain small size were not worth collecting, and passed them by. Most of these were picked up by others, and therefore found their way into the collection. On another occasion a team member got tired of finding only L-group chondrites and decided that we had already found enough of them. I was not with either of these field teams, but I am told the problems were speedily corrected. These events would have had minimal effect, in view of the thousands of specimens that have been collected.

One potential source of *non-personal* bias affected all team members equally. Some meteorite sites were located around the edges of moraines where terrestrial rocks were much more abundant than meteorite fragments. Some meteorites are closely similar in appearance to terrestrial rocks, so some undoubtedly were missed. Now and then a terrestrial rock finds its way back to Houston labeled as a meteorite, but these are very few. The fact that misidentification of terrestrial rocks as meteorites occurs only rarely may indicate that misidentification of true meteorites as terrestrial rocks also occurs infrequently.

Nevertheless, "all else being equal..." will not go away, and we must live with it. For now, I will assume that all sources of bias in the collection have been accounted for. We still have to decide if we have a large enough collection to be a representative sample in all its parts (Hint: we don't).

Mass–frequency distributions

It would be nice to have some criterion to use in determining whether or not we had collected large enough samples of each meteorite type to be representative of the entire population of meteorites on the ice. One way to do this would be to wait 30 years until the antarctic collection presumably will have doubled in size, and see if the new, revised Table 9.3 still gave the same relative proportions of stones/irons/stony irons. It is unlikely, however, that I have 30 years left, and anyway, I wanted to know *now*. So I decided to look at *mass–frequency distributions*.

The ANSMET collection gives us an opportunity not available with the modern falls: we can compare recovered masses to numbers of finds. This is important, because we can construct mass–frequency distributions for the specimens making up a group. If such a curve is unimodal (i.e., tends toward having a single peak) we can have a greater degree of confidence that we have an adequate sample size. Bimodality, or a double peak, could indicate that we have sampled two different populations or that we do not have enough specimens in the sample to be representative of the population from which it came.

As an example, suppose we were to summarize the weights of 1.8-m tall American males. We could add up the measured weights of 1000 such men and divide by 1000 to get the weight of the average 1.8-m tall adult American male. But we would find that not every man is average: some weigh more than the average and some weigh less. In our sample of 1000, we would find that there would be many individuals who were slightly above the average, but successively fewer overweight individuals as the degree of obesity increased. On the other side of the average, there would be many underweight individuals who were close to the average, but successively fewer underweight individuals as the deviation from the average value tended toward the fatally anorexic[4]. If we sorted the data from our sample of 1000 men into 1-kg bins and plotted the numbers in each bin, we would get a curve describing the frequency of weight distribution among 1.8-m tall adult American men. This is a *mass–frequency distribution* (better known as a frequency polygon, but that is statisticians' jargon). There is a good chance, also, that this curve would be symmetrical about the mean; in other words, the curve on one side of the average would be a mirror image of the curve on the other side, and the average value would

---

[4] The average does not define the healthiest weight: it is generally conceded that the average American male is substantially overweight. The data simply describe the world as it is, not as it should be.

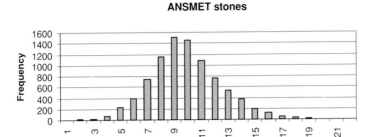

FIGURE 9.3 Mass–frequency distribution curve for 8722 ANSMET stones. This is a good example of a unimodal curve that characterizes a sample of adequate size.

occur with the highest frequency. I have described a bell-shaped, or *normal*, curve.

Suppose now that we wish to construct a frequency distribution for the weights of adult American males who are 2.1-m tall, but we can conveniently find only 10 men who are that tall. This is a much smaller sample. We can plot a mass–frequency distribution using these measurements, but there would be gaps in the data that could suggest bimodality and we could not say with any degree of certainty that we are adequately describing variations within that population of men[5]. Certainly we could not assign great validity to a comparison of the frequency distributions for the one thousand 1.8-m tall men with that for the ten 2.1-m tall men. Statisticians describe the latter as an inadequate sample. We shall see that some meteorite types and groups are too few in number to provide an adequate sample. Figures 9.3–9.5 are the mass–frequency distributions for the ANSMET meteorite types (stones, irons and stony irons) listed in Table 9.3.

Based on Figures 9.3–9.5, we can conclude that sample sizes for irons and stony irons in the ANSMET collection are not yet large

---

[5] Also, they would all be basketball players, but to keep it simple let us assume there are no other sources of bias.

ANSMET irons

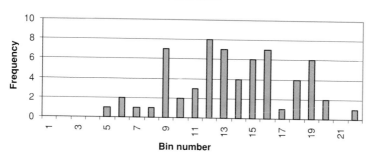

FIGURE 9.4 Mass–frequency distribution for 63 ANSMET irons. The total number of irons plotted here differs from that in Table 9.3 because the Derrick Peak and Inland Forts irons are included here. Note that even with 63 data points, the frequency distribution for irons cannot be characterized as unimodal and, while the number represents the true proportion of irons in the collection, it might be risky to assume it represents the true proportion of irons lying on the ice.

ANSMET stony irons

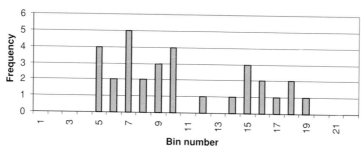

FIGURE 9.5 Mass–frequency distribution for 31 ANSMET stony irons. There are only 4 pallasites and 27 mesosiderites in the ANSMET collection, so this is essentially a distribution for mesosiderites. This is another example of a curve produced by a sample of inadequate size. Note also that if any inference can be made about this distribution it is that it could be bimodal. If that were the case, it would suggest that the distribution of mesosiderites on the ice represents two different populations, each of which has been inadequately sampled.

enough to reliably represent their proportions on the ice in Antarctica. This suggests that any attempt to compare stones/irons/stony irons between the ANSMET collection and the modern falls would be risky, particularly since the numbers for irons and stony irons in the

modern falls collection are equally small. There are statistical procedures that could be employed to estimate reliability of the data. The
Kolmogorov–Smirnoff and chi-squared tests have been shown to be
useful but these are perhaps better investigated in a dedicated scientific paper.

Skewed curves

Here is something I have not told you yet. The horizontal scale used in
Figures 9.3–9.5 is a logarithmic scale. Aside from the sound of more
books hitting the wall, what does this mean? Table 9.5 shows the
mass ranges corresponding to each bin. Each bin accepts twice the
mass range of the one before it.

While the curve of Figure 9.3 has a unimodal shape, leading us
to accept it as a sample of sufficient size to be representative of the
stones lying on the ice, it is not a normal (bell-shaped) curve. That
it looks *almost* like a normal curve is an artifact of the horizontal
scale, which is not linear. In Figures 9.3–9.5, individual meteorite
masses range between 0 and 131 072 g. In order to compress this range
enough for presentation in a convenient space, each succeeding bin
size is twice as large as the one preceding it[6] (i.e., a log scale). Bin 5,
for example, contains all specimens between 0.5 and 1 g, while bin 6
contains all specimens between 1 and 2 g. Bin 7 contains all specimens
between 2 and 4 g, and so on. The same set of data plotted on a linear
scale would tail off very far to the right of the highest peak in the
curve, and the skewed nature of the curve would be easy to see.

Remember that in the data for 2-m tall men, which had a linear
scale with every bin size equal to all others, the average weight (the
mean) was also the value that occurred most frequently (the mean
had the same value as the mode). In the curve for ANSMET stones,
the average mass of 8722 specimens is 216.6 g. This mass is found
in Bin 13 (128–256 g), but the most frequently occurring size, which
statisticians like to call the *mode*, is in Bin 9 (8–16 g). When the average

---

[6] An exception is Bin 1, which contains all masses between 0 and 0.0625 g.

Table 9.5. *The mass range in each bin*

Bin number	Mass range (g)	
	From	To
1	0.000	0.062
2	0.063	0.125
3	0.125	0.250
4	0.250	0.500
5	0.500	1.000
6	1.000	2.000
7	2.000	4.000
8	4.000	8.000
9	8.000	16.000
10	16.000	32.000
11	32.000	64.000
12	64.000	128.000
13	128.000	256.000
14	256.000	512.000
15	512.000	1024.000
16	1024.000	2048.000
17	2048.000	4096.000
18	4096.000	8192.000
19	8192.000	16384.000
20	16384.000	32768.000
21	32768.000	65536.000
22	65536.000	131072.000

(mean) value differs from the mode, the curve is no longer normal – it is *skewed*. That this is a skewed distribution curve, with the most common size being quite small, has significance for the ANSMET field team in that the stony meteorites they will find most frequently are about the size of a large olive – this is less exciting than finding a big meteorite. But it is significant also in a more fundamental sense, because this type of *skewed* frequency distribution is very much like the frequency distributions observed for rock fragments that have been produced by a rock crusher. This similarity was already pointed out by Ralph Harvey, who is now the principal investigator of the ANSMET project, in his PhD thesis at the University of Pittsburgh. Because of this similarity, we can suspect that stony meteorites have suffered the effects of physical crushing processes. It is easy to accept such a conclusion for the following reasons:

- We assume that fragment-producing collisions occurred commonly between bodies in the asteroid belt.
- When a meteorite enters the earth's atmosphere at hypersonic velocity, it accumulates a wedge of highly compressed gas in front of it. This wedge of gas transmits a back pressure against the face of the meteorite that it may not be strong enough to resist, so it breaks apart.
- When the meteorite strikes the surface it may fracture.
- Burial at depth in the moving antarctic ice sheet may fracture a meteorite (but it must be said that the few we have found still embedded in ice have not supported this concept).
- Once exposed on the ice surface[7], chemical weathering is accelerated, leading to slow disaggregation. The results can resemble physical crushing.

For all these reasons, it is not surprising that the frequency distribution curve for stony meteorites is so similar to that for fragments of rocks that have been through a crusher. Recognition of this, however, need not deter us from accepting a unimodal (albeit skewed) mass distribution curve as a representative sample.

The numbers for the ANSMET stony meteorite finds (Table 9.3), inspire confidence that the sample from which they are derived is large enough to represent their entire population on the ice. This is supported by their mass–frequency distribution (Figure 9.3). But the

---

[7] See the following chapter for a discussion of this process.

number of stony meteorites in the *modern falls* is also large enough to inspire confidence that it is an adequate sample of stony meteorites falling today. In fact, if the 8643 stones in the ANSMET collection averaged 10 members to a fall, the number of antarctic ANSMET falls would be virtually identical to the numbers for modern falls. This would be a case of meteorites collected in a small area over a great length of time (the antarctic collection) being equivalent to meteorites collected over a very large area during a short length of time (the modern falls).

Chondrites vs. achondrites

Because the overall curve for stony meteorites (which includes the subtypes chondrites and achondrites) is nicely unimodal, we can have some confidence that we have a representative sample of the population of stony meteorites on the ice in Antarctica. Considering the population of stony meteorites in isolation, then, can we examine the relative abundances among subsamples of this population? The hope would be that each subsample will contain large enough numbers to give reliable relative abundances. Tables 9.6 and 9.7 give data for the chondrite and achondrite subtypes. These are the only two subtypes of stony meteorites. Each consists of a number of groups.

The two subtypes of stony meteorites are chondrites and achondrites. Among the modern falls, there are 781 chondrites and 73 achondrites that have been identified as to group. Among the ANSMET finds, the collection contains 8409 chondrites and 234 achondrites. The mass–frequency distribution for the chondrites is nicely unimodal (Figure 9.6) but the curve for 234 achondrites is a little less confidence-inspiring (Figure 9.7).

We might be able to use the achondrite data shown in Figure 9.7, but there is a better alternative. I remarked earlier that because of the sampling philosophy followed by the Japanese scientists, only the rarer groups in their collection were selected for characterization, while many of the remaining stones were labeled "stony meteorite," or are not labeled at all. Also, we can assume that most of the

Table 9.6. *Chondrites and achondrites among the modern falls and ANSMET finds. The ANSMET finds numbers are slightly different from the corresponding numbers in Tables 9.1 and 9.3 because ungrouped stones are not included here. This also causes a slight difference in total mass*

Subtypes and groups	Modern falls N	ANSMET finds	
		N	Mass(g)
Chondrites			
EH	8	53	1730
EL	7	21	3104.4
H	314	3545	608113.4
L	342	3914	1156542.4
LL	72	646	41216.9
R	1	7	625.1
K	1	1	23.1
CI	5	0	0
CM2	15	70	6947.5
CO3	5	9	2566.2
CV3	6	20	1114.1
CK	2	61	3056.6
CR2	3	55	1537.8
CH	0	7	99.9
Subtotals	781	8409	1826677.4
Achondrites			
BRACH	0	1	4.6
ACAP/LOD	2	12	1150.5
URE	5	44	10907.5
AUB	9	35	4009.8
HOW/EUC/DIO	55	139	34289.6
ANGR	1	2	7.5
WIN	1	1	11.3
Subtotals	73	234	50380.8
Totals	854	8643	1877058.2

Table 9.7. *Percentage abundances of chondrites and achondrites among the modern falls and the ANSMET finds, derived from Table 9.6*

Subtypes and groups	Modern falls N(%)	ANSMET finds	
		N(%)	Mass(%)
Chondrites			
EH	0.9	0.6	0.1
EL	0.8	0.2	0.2
H	36.8	41.0	32.4
L	40.0	45.3	61.6
LL	8.4	7.5	2.2
R	0.1	0.1	0.0
K	0.1	0.0	0.0
CI	0.6	0.0	0.0
CM2	1.8	0.8	0.4
CO3	0.6	0.1	0.1
CV3	0.7	0.2	0.1
CK	0.2	0.7	0.2
CR2	0.4	0.6	0.1
CH	0.0	0.1	0.0
Subtotals	91.5	97.3	97.3
Achondrites			
BRACH	0.0	0.0	0.0
ACAP/LOD	0.2	0.1	0.1
URE	0.6	0.5	0.6
AUB	1.1	0.4	0.2
HOW/EUC/DIO	6.4	1.6	1.8
ANGR	0.1	0.0	0.0
WIN	0.1	0.0	0.0
Subtotals	8.5	2.7	2.7
Totals	100.0	100.0	100.0

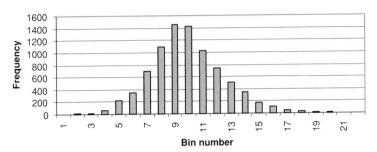

FIGURE 9.6 Mass–frequency distribution of 8409 ANSMET chondrites.

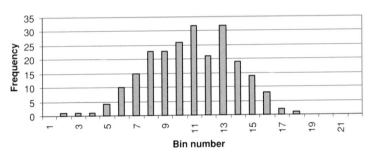

FIGURE 9.7 Mass–frequency distribution of 234 ANSMET achondrites.

EUROMET achondrites have been identified. Even if not, it wouldn't make much difference to the statistics. So if we assume that all *achondrites* in all the antarctic collections have been correctly identified, then the remaining stones must be chondrites. The total of all antarctic collections then contains 17 249 chondrites (Figure 9.8) and 431 achondrites (Figure 9.9).

Notice, in comparing Figures 9.6 and 9.8, that while the numbers have doubled, the antarctic chondrite distribution looks much like its smaller ANSMET component. In comparing Figures 9.7 and 9.9, however, there are differences. In Figure 9.7, the curve lacks the degree of skewness we have come to expect. In Figure 9.9, the mode can be identified more easily and the curve is skewed much further toward smaller masses. In short, it has more of the characteristics we

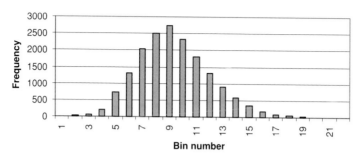

FIGURE 9.8 Mass–frequency distribution for 17249 antarctic chondrites.

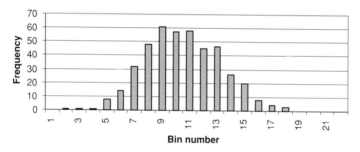

FIGURE 9.9 Mass–frequency distribution for 431 antarctic achondrites.

expect in mass–frequency distribution curves of antarctic meteorites. Table 9.8 presents the data on numbers of modern fall chondrites and achondrites, and numbers and masses of these same subtypes in the *total antarctic collection*. Table 9.9 gives the percentage abundances calculated from the data.

We can see in Table 9.9 that within the antarctic collection the number percentages and the mass percentages of chondrites vs. achondrites give almost equal relative abundances of these subtypes (among stony meteorites, density differences are not very significant). These relative abundances, however, are very different from those abundances in the modern falls. Because the mass–frequency distribution curves suggest that the ANSMET samples are large enough to be representative samples, we can suspect that we are looking at a

Table 9.8. *Data for chondrite vs. achondrite abundances in the modern falls and the* total antarctic collection

Subtypes	Modern falls N	Antarctic finds N	Antarctic finds Mass (g)
Chondrites	781	17249	2854038
Achondrites	73	431	78894
Totals	854	17680	2932932

Table 9.9. *Percentage abundances of chondrites vs. achondrites. Abundances by numbers in the modern falls and abundances by numbers of specimens and by mass in the* total antarctic collection

Subtypes	Modern falls N(%)	Antarctic finds N(%)	Antarctic finds Mass (%)
Chondrites	91.5	97.5	97.3
Achondrites	8.5	2.5	2.7
Totals	100.0	100.0	100.0

real, quantitative difference between the two collections. Table 9.7, which gives relative abundances for the groups within chondrites and achondrites shows that almost all of this difference can be accounted for among the HED achondrites. Since the modern falls represent data accumulated over about the last 200 years and the antarctic collection records data on meteorites that have fallen over several hundred thousand years, the conclusion seems pretty clear that anomalously large numbers of HED meteorites are reaching the earth today.

How can this happen, when we know that evolutionary changes in the flux of meteorites reaching Earth require one or two million years, at least, to occur by resonance-related perturbations in the asteroid belt? Well, we need another mechanism of delivery with a shorter response time.

In 1991, Dale Cruikshank, at NASA Ames Research Center, Moffett Field, California, and some colleagues, succeeded in measuring the reflection spectra of three known earth-approaching asteroids. They found some remarkable similarities between these spectra, the asteroid Vesta and HED meteorites. It is commonplace to suspect that the HED meteorites originated as part of Vesta, but they suggested that perhaps these meteorites actually came, most recently at least, from one or more of the Vesta-like asteroids that already spend a significant fraction of their time in our vicinity. (Whether these asteroids really derive from Vesta, or not, is another question.)

The Vesta-like earth-approaching asteroids are named 3551, 3908 and 4055. Their diameters range between 0.5 and 3 km. These bodies have points of closest approach to the sun (perihelia) just outside the orbit of the earth and points of farthest retreat from the sun (aphelia) within the asteroid belt. As a matter of some interest, we do not expect them to collide with the earth unless they suffer future orbital changes that make them earth-crossers. If that happened, we should start to worry because such large individuals could cause considerable damage if they survived intact during transit through the atmosphere. Chances are, however, that they would break up and fall in a giant shower of smaller fragments. (So please try not to worry.)

Getting back to the present situation, each of these asteroids makes a round trip out to the asteroid belt and back to the earth's vicinity every 2.5–3 years, and they each spend some time in the asteroid belt on every round trip. One could suspect that one of them recently suffered a glancing collision with another body in the asteroid belt that might have removed 100 to 1000 metric tons (tonnes) of surface material (just a guess about the quantity). Such a collision could have been gentle, by asteroid-belt standards: maybe only a touch from another body that was revolving in a different sense, so that a lot of surface material was scraped off. The newly liberated meteoroids would have been placed into slightly different orbits than their parent asteroid. Those that received orbital velocities slightly less than their parent would be moving slower at aphelion. Their perihelion distances would decrease and they would become earth-crossing bodies, subject to possible collision with the earth. Because of their very short orbital periods they would be earth-crossers every few years, and the likelihood of colliding with the earth would be relatively great. This mechanism for altering the meteoroid flux in the vicinity of the earth does not rely at all upon gravitational perturbations by Jupiter, and the results would become noticeable over a very short time, of the order of hundreds to thousands of years. *So maybe we should suspect that the flux of HED achondrites reaching the earth has increased very recently, and maybe this entire exercise has been worthwhile. (Ho, ho! Hoo, hah!)*

H, L and LL chondrites

We have seen that the antarctic meteorite collection can suggest something interesting about the HED meteorites. The only other meteorite groups that have large enough proportions in the collection to be considered statistically significant are the ordinary chondrites. Because many ordinary chondrites are not specifically identified in the Japanese collection, and the EUROMET and Italian collections are relatively small, I choose to rely on their numbers in the ANSMET collection. Their mass–frequency distributions are given in Figures 9.10–9.12.

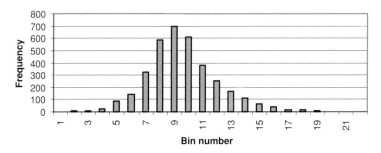

FIGURE 9.10 Mass–frequency distribution for 3545 ANSMET H group chondrites.

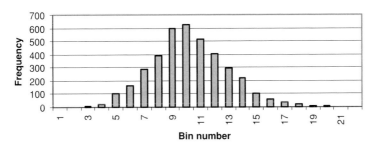

FIGURE 9.11 Mass–frequency distribution for 3881 ANSMET L group chondrites.

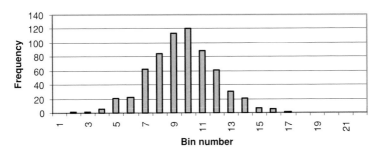

FIGURE 9.12 Mass–frequency distribution for 646 ANSMET LL group chondrites.

The fact that all these mass–frequency distributions are unimodal suggests that the numbers leading to each are large enough to represent reliable samples of the populations of these meteorite groups on the ice in Antarctica. If the average number of fragments from the fall of these meteorites in Antarctica were 10, then the numbers of falls of ordinary chondrites in the ANSMET collection would be about the same as the numbers represented in the modern falls collection. This is a suggestion that the numbers in the modern falls collection are reliable indicators of the relative proportions of H/L/LL chondrites falling today, and we could conceivably learn something by comparing their abundances.

Table 9.10. *Data for H, L and LL chondrite abundances in the modern falls and the ANSMET collection*

	Modern falls	ANSMET finds	
Groups	*N*	*N*	MASS (g)
H	314	3545	608113
L	342	3881	1156542
LL	72	646	41217
Totals	728	8072	1805872

Table 9.11. *Percentage abundances of H, L and LL chondrites. Abundances by numbers in the modern falls and abundances by numbers of specimens and by mass in the ANSMET collection*

	Modern falls	ANSMET finds	
Groups	*N*(%)	*N*(%)	MASS(%)
H	43.1	43.9	33.7
L	47.0	48.1	64.0
LL	9.9	8.0	2.3
Totals	100.0	100.0	100.0

Tables 9.10 and 9.11 give the data for the modern falls and the ANSMET collection and the percent abundances of H, L and LL chondrites.

In Table 9.11, compare the percent abundances by mass with percent abundances by numbers of specimens in the ANSMET collection. It is easy to verify that they are quite different. The reason for this apparently involves average specimen size. H chondrite specimens have an average size of 171.5 g. Average sizes of L chondrites and LL chondrites, respectively, are 298 g and 63.8 g. A given number of L chondrites on the ice in Antarctica would weigh 1.75 times as much as the same number of H chondrites and 4.67 times as much as the same number of LL chondrites. So while the number abundances would be 33.3%:33.3%:33.3%, the mass abundances would be 32%:55.9%:12.0% for the H:L:LL groups. *So we have discovered the second reason that abundances by mass may not correspond to abundances by numbers – the average mass may be different for different groups.* (You will remember that density differences between types provided the first reason.) Notice that while abundances by mass and abundances by number in the ANSMET collection are not the same, abundances by numbers in the ANSMET collection do correspond pretty closely to abundances by numbers in the modern falls collection. So we can immediately sense a difference here between H, L, and LL chondrite groups on the one hand, and HED achondrites on the other. This is that contrary to the case of the HED meteorites, recent changes have not occurred in the flux of H, L and LL chondrites falling to Earth. Thus, special mechanisms do not have to be invoked to get any of the ordinary chondrites from the asteroid belt to Earth, and we can assume they arrive through the generally accepted mechanism of transfer to resonance frequencies and gravitational perturbation by Jupiter.

CAN ANTARCTIC METEORITES TELL US ANYTHING ABOUT THE ASTEROID BELT AND ITS HISTORY THAT WE WOULD NOT HAVE KNOWN WITHOUT THEM?

The short answer is, "Probably not." But they can give us a more quantitative approach to some old, and some new, speculations.

A game called pairing

Before we had antarctic meteorites, when estimates of the abundances of different meteorite groups were arrived at by tabulating the relative numbers of falls, it was important not to count a fall more than once. It was recognized that many falls consisted of multiple fragments that fell relatively close to one another. These were said to be "paired," or "fall-paired," and were counted as only one fall.

When we began to look at lunar meteorites and wonder how many sites on the moon these meteorites represented we began to speak of "site pairing." Site-paired lunar meteorites did not have to have fallen at the same time or onto the same part of the earth, they just had to have a (rather general) petrological similarity and a common age of ejection to establish the likelihood that they came from the same impact event and sampled the same small part of the moon.

But lunar meteorites and, of course, martian meteorites are also paired in another sense – as coming from a specific parent body. They have been parent-body-paired. In the classification of asteroidal meteorites, this is the function of the meteorite groups listed in Table 9.1. Groups are considered to contain all specimens that are "parent-body-paired" and we speak, for example, of the H-group parent body, or the angrite parent body. For this reason it is not important to know how many falls the antarctic meteorites represent, which we cannot know anyway. We can do parent-body pairing, which is intrinsically more interesting.

Looking back now at the list of stony meteorites in Table 9.1, we can understand that petrologists have assigned them to 21 named groups because they are different enough from each other to have come from 21 different parent bodies. Some of the "ungrouped" stony meteorites probably come from parent bodies other than these 21 but, because we have only one or two specimens of each, we cannot yet characterize them as separate groups.

The same considerations apply to irons and stony irons, except that because iron meteorites exist we assume some asteroidal parent bodies were large enough to have formed nickel-iron cores. Cores must have mantles, therefore some groups of irons, and the two groups of stony irons, may be paired with some groups of stony meteorites. Even with potential parent-body-pairings between meteorite types, however, it still seems fairly certain that at one time there were many meteorite parent bodies in existence. Current thinking places most of them, if not all, in the asteroid belt.

Among the ways that different meteorite groups, and therefore their parent bodies, differ from each other is in bulk composition, so we can think of the asteroid belt, where the parent bodies of meteorites were spawned, as a very wide zone that varied somewhat in composition across its width, or from place to place.

Perspectives on the rarer groups of meteorites

Most papers at any scientific meeting devoted to meteorites deal with the rarer groups. The reason is that each new member of a rare group has the potential to tell us things about that group (and its parent body) that we didn't already know. By contrast, it is a very unusual paper that gives new insights into the parent bodies of ordinary chondrites, just because so many of them already have been examined. Because we are used to hearing more about the rarer specimens it may be surprising to some to notice just how rare they are, as shown in Tables 9.1 and 9.2. Table 9.1 demonstrates that the antarctic meteorite collection has made significant contributions to the "rare" category of meteorite groups. While we cannot compare the relative abundances of these groups, because the total numbers are too small, we can claim that these meteorites are falling only at an extremely low rate, and have been doing so for at least the time period covered by the terrestrial ages of the bulk of antarctic meteorites. Just as the abundant ordinary chondrites must reflect events in the asteroid belt, so must these rare meteorites.

Some formerly separate meteorite groups have been parent-body-paired. Examples are the howardites, eucrites and diogenites (the HED meteorites). Some of these probably come from one or more of the Vesta-like earth-approaching asteroids which, in turn, may have come from Vesta. The evidence is that light reflected from all these objects gives a very similar spectral signature. In addition, howardites appear to be physical mixtures of eucrite fragments and diogenite fragments. Acapulcoites and lodranites were earlier classified in different groups, based upon compositional and textural differences. Today they are connected by a couple of antarctic specimens that are intermediate to them in composition, and this suggests that these two groups are from a common parent body. Some of the iron meteorite groups contain inclusions of known stony meteorite groups, suggesting a connection between meteorite types in some differentiated parent bodies. Every such apparent connection that can be established helps us to build a picture of the composition and structure of parent bodies that existed in the asteroid belt at some time in the distant past. The rarer meteorite groups are just as important in this endeavor as the ordinary chondrites.

Why are some meteorite groups so rare?
Another way of asking this question may be, "Why are ordinary chondrites so common?" It may be considered unfortunate, by those who like unequivocal answers, to learn that there are at least three possible answers that would satisfy both questions.

(1) The relative rarity of a meteorite group can be taken to reflect the abundances of parent-body materials that were available in the asteroid belt. The materials leading to ordinary chondrites were just more abundant there. As a result, most of the parent bodies that formed in the asteroid belt had compositions typical of H, L and LL chondrites.

(2) Dynamical studies of the fate of objects in the asteroid belt indicate that bodies occupying certain zones will drift more readily into resonance orbits and find their way into the meteoroid flux at the earth more frequently. The relatively high abundance of H, L and LL chondrites in our collections simply reflects the fact that the parent bodies of these meteorite groups occupied zones that were closer to resonance zones.

(3) The flux of meteoroids in the earth's vicinity has changed in character over time, reflecting variations in the relative supply of fragments from different meteorite groups available for resonance-related perturbations. Ordinary chondrites represent the most recent wave of flotsam to wash up on our shores.

We have a complicated situation here, because in addition to three different answers to the same set of questions, each answer may be partly true, and the proportion of "partly" that is assigned to each answer may vary from worker to worker.

At this point we have entered an arena where arguments are highly speculative, but most speculations will swirl around the relative importance of Answers 1 and 2. Answer 3 may be a new one. I think I thought of it alone, and unaided, but so little is really new these days that I cannot feel safe in making that claim[8]. What I propose to do is try to find support for Answer 3 in a further examination of the data on the ANSMET collection and to bring in supporting arguments from observations on meteorites in general. My hope is to validate Answer 3 as a basis for discussion. The reader should feel free to form his/her own opinions. The thesis will be as follows.

The history of the asteroid belt is one of collisions between parent bodies. Some parent bodies were destroyed before others, and some survived, by chance, for quite a long time. When a parent body is destroyed it produces clouds of fragments. Some of the fragments will be thrust immediately into resonance orbits and will quickly leave the asteroid belt. Others will drift into these regions over time. With the passage of time the ranks of any particular group of parent-body fragments remaining in the asteroid belt will inevitably and progressively thin.

[8] As an example, in a 1990 paper in which he discussed the apparent overabundance of ungrouped irons in the antarctic collection, John Wasson, at UCLA, pointed out that most of these were smaller-than-average specimens. He suggested that a fragment-producing impact in the asteroid belt would impart a greater velocity to small-mass specimens than to fragments of larger mass. This would result in a greater variety of orbital parameters for smaller fragments and make more of them available to resonance orbits. Also, subsequent jostlings the fragment may have received would have the same effect, and the more jostlings, the greater the effect. This could be taken to imply that the longer the fragment had been available for jostling the earlier had its parent body been destroyed. On the other hand, it could as easily be taken to imply that the parent body was just located in a region of the asteroid belt from which it is difficult to supply material to the resonance orbits, i.e., Answer 2 above. John does not say which of these ideas he preferred at the time, but his paper is certainly thought-provoking. In the untraceable interplay between deduction and synthesis that occurs during the creative process, John's paper may ultimately have led to my formulation of the more generalized Answer 3, above.

Theoretically, a five-year-old automobile still has some molecules of fuel in its gas tank that were present in its first full tank of gas. A 10-year-old car has fewer of these original molecules, but some still remain. By analogy, fragments of all parent bodies still exist in the asteroid belt. One could believe that the rarer meteorite types in our collections may have had a longer history of depletion in the asteroid belt and were once much more common messengers to the earth.

The low abundances of the rarer meteorite groups in our collections suggests something interesting about the overwhelmingly common ordinary chondrites. This is that the parent bodies of the H, L and LL chondrites were among the last to be destroyed. We can infer this because we are still receiving them so commonly.

Two ways that parent bodies can be destroyed
We could imagine a mutual collision between two bodies of roughly equal size that is so energetic that both are reduced to fragments traveling away from the collision point in all directions (see Figure 9.1). We could as easily imagine a collision between two bodies of unequal size in which the smaller body is destroyed but the larger only receives surface scars, and a slight diminution in mass. A variant on this could be a glancing collision between two large bodies that removes a little surface material from both. By one method, destruction is cataclysmic. By the other methods, if repeated frequently enough, eventual destruction also is assured.

Early in solar system history, when the spatial density of parent bodies was high, mutually disruptive collisions may have been more common. Later, when there were fewer large bodies and many, many more small ones, parent bodies might more commonly die by gradual attrition, from the surface layers, inward. If the ordinary chondrite parent bodies have been among the last to be destroyed, one would expect that they have died a slow death, by attrition. In that case, it could be interesting to look at the relative abundances of their metamorphic grades currently reaching the earth.

Metamorphic grade among the ordinary chondrites

If the ordinary chondrites in the ANSMET collection are numerous enough to give unimodal mass distribution curves, perhaps subgroups within the H, L and LL groups also have large enough abundances to be considered representative of their proportions on the ice in Antarctica. Ordinary chondrites are divided into well understood subgroups based upon metamorphic grade. Data on metamorphic grades among the ordinary chondrites are included in Table 9.12 and relative proportions of their metamorphic grade subgroups are found in Table 9.13.

The mass–frequency distributions by metamorphic grade for ANSMET LL, H and L chondrites are shown in Figures 9.13–9.15.

In breaking the chondrites into groups and then subdividing the groups according to metamorphic grade, we are trying to compare ever-smaller clusters of objects, and at some point we must decide that the numbers are not large enough to truly represent the distribution of subgroups on the ice. Inspection of the mass–frequency distributions in Figures 9.13–9.15 reveals that some of them are definitely ragged, and less reliable in appearance than we might prefer. Metamorphic grades 3 and 4 in each group are particularly suspect, and any quantitative comparison of them with modern falls could be meaningless.

But there is a point to be made here. In a qualitative sense, grades 3 and 4 for all the ordinary chondrites are less abundant among both the modern falls and the ANSMET finds. In a decent sized parent body, one would expect the lower metamorphic grades to be located nearer the outer surface and the higher metamorphic grades more toward the center. If these parent bodies were diminishing by attrition, the outer portions would be the first to go. They would also be the most abundant fragments from each parent body to arrive first at the earth and the first to become rare as their source diminished and contributions from deeper layers began to arrive in increasing numbers. We could make the following argument: *if* the parents of the ordinary chondrites have been among the most recent parent bodies to be destroyed, and *if* they have been destroyed by attrition, then the relative abundances

Table 9.12. *Numbers of modern falls and numbers of specimens and their masses in the ANSMET finds collection*

Groups and subgroups	Modern falls N	ANSMET ordinary chondrites		
		N	Mass (g)	Av. mass (g)
H chondrites				
H3	5	37	6380	172.4
H4	57	274	117669	429.5
H5	138	2166	346920	160.2
H6	92	1054	134178	127.3
Totals	292	3531	605147	N.A.
L chondrites				
L3	7	213	42236	198.3
L4	20	155	29574	190.8
L5	61	842	162128	192.6
L6	239	2644	914054	342.5
Totals	327	3854	1147992	N.A.
LL chondrites				
LL3	9	33	4564	138.3
LL4	8	18	3164	175.6
LL5	15	371	15539	41.9
LL6	35	221	17920	81.1
Totals	67	643	41187	N.A.

Av., average; N.A., not available.

we think we see among the metamorphic grades might be not very surprising.

Given the recent history in the asteroid belt that I have been suggesting, one other facet of the ANSMET data could be discussed. Notice that I included a column in Table 9.12 labeled "Average Mass." I included this column to show that, even though there are no extreme density differences between H, L and LL chondrites that would explain

Table 9.13. *Percent abundances derived from Table 9.12*

Groups and subgroups	Modern falls N(%)	ANSMET ordinary chondrites	
		N(%)	Mass(%)
H chondrites			
H3	1.7	1.0	1.1
H4	19.5	7.8	19.4
H5	47.3	61.3	57.3
H6	31.5	29.8	22.2
Totals	100.0	100.0	100.0
L chondrites			
L3	2.1	5.5	3.7
L4	6.1	4.0	2.6
L5	18.7	21.8	14.1
L6	73.1	68.6	79.6
Totals	100.0	100.0	100.0
LL chondrites			
LL3	13.4	5.1	11.1
LL4	11.9	2.8	7.7
LL5	22.4	57.7	37.7
LL6	52.2	34.5	43.5
Totals	100.0	100.0	100.0

differences between the number abundances and mass abundances of Table 9.13, there are differences in average masses. We should not expect correspondence between number abundances and mass abundances, and, indeed, we do not see it. In this case, therefore, I prefer to discuss abundances based on average mass because I believe that mass is a more fundamental quantity than number. This has the drawback that we cannot compare results to the modern falls, but we will labor on, nevertheless.

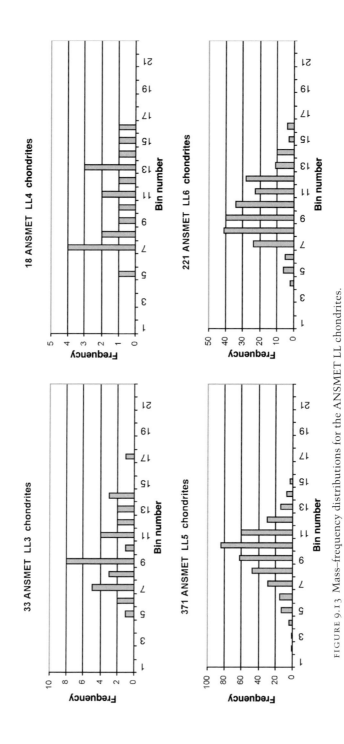

FIGURE 9.13 Mass–frequency distributions for the ANSMET LL chondrites.

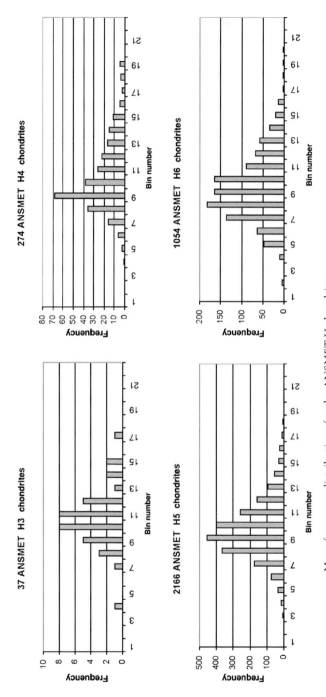

FIGURE 9.14 Mass–frequency distributions for the ANSMET H chondrites.

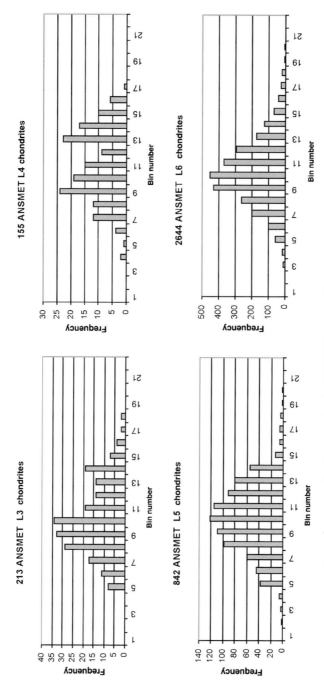

FIGURE 9.15 Mass-frequency distributions for the ANSMET L chondrites.

Laboring on, then, we can compare the mass abundance ratios H5/H6, L5/L6 and LL5/LL6. We can see that H5/H6 >1 but L5/L6 and LL5/LL6 are both <1. Since the numbers of specimens in each of these subgroups are large and the mass–frequency diagrams look uni-modal, we may be looking at real abundances on the ice and thus in the neighborhood of Earth. If so, one could conclude that we are see-ing a stage in the destruction of the H parent body that is slightly less advanced than that of the other two, because H5 material is still falling in greater relative amounts than H6. Also, L5/L6<<1, while LL5/LL6 is close to one. So, of the three parent bodies of or-dinary chondrites, we are sampling the most advanced stage of de-struction with L chondrites, a lesser stage with LL chondrites and the least advanced stage with H chondrites. Some will say that we reached the practical limits of speculation well before we reached this point.

How important has gradual attrition been as a mode of destruction?

Aside from the observations we made above, there are other clues that can be cited that do not contradict gradual attrition of a parent body during tens of millions or a couple of billion years as the normal mode of destruction in the asteroid belt. As described in Chapter 8, the achondrite groups ureilite and aubrite, which are among the rarer groups of stones, apparently are products of igneous fractionation. The fractionation process would have separated ureilite and aubrite accu-mulations from an overlying magma of different composition. The magma, once separated, would have crystallized to rocks of differ-ent composition than the underlying aubrites and ureilites. But the implied products of these fractionation processes, the overlying rocks, are not found in our meteorite collections. Others have suggested that these missing layers have been destroyed. If these outer layers of their parent bodies had been removed earlier by attrition, the outer-layer fragments would have been processed earlier out of the asteroid belt and depleted earlier from the flux reaching Earth. Even the underlying ureilite and aubrite compositions are now very rare among meteorites reaching the earth, suggesting that remnants of the material that was removed earlier are even more rare.

Vesta is probably the best example that can be cited. The three Vesta-like earth-approaching asteroids and the HED achondrites all may have come ultimately from Vesta. So in Vesta we may be seeing the attrition process in action.

It seems to me, too, that attrition could act much more effectively on the stony mantles of differentiated parent bodies than on their nickel-iron cores. Over time, destruction of differentiated asteroids by attrition could produce clouds of stony meteoroids but only a few very large nickel-iron masses. The abundance of stony fragments would ensure that many more small fragments could serve as feedstock for the resonance zones, but only the occasional large iron would wander in. As a result, there may be a disproportionate number of relatively large nickel-iron parent-body cores still muscling their way around the asteroid belt and occasionally nudging what remains of an H, L or LL parent body, or Vesta, with deleterious results, *particularly for the nudgee.* This could also explain how the flux of HED achondrites was increased without a corresponding increase in fragments of the other body involved in that collision. (This could have been a problem with the HED story, but I conveniently failed to mention it at the time.)

I should emphasize to the general reader that interpretations I may make, based on the data in this chapter and other chapters, are speculations. There are alternative interpretations for all of the observations, and it is certain that some of mine are wrong. One could speculate endlessly without reaching any solid conclusions. But speculation is fun, and can be thought-provoking. What makes these speculations possible is the existence of the antarctic meteorite collection and the fact that we have both masses and numbers for the specimens in this collection.

SUGGESTED READING

A nice review of studies involving Vesta, vestoid asteroids and HED meteorites is given in:

Drake, M.J. (2001) The eucrite/Vesta story. *Meteoritics and Planetary Science* **36,** 501–513.

10 Meteorite stranding surfaces and the ice sheet[1]

METEORITES AND BLUE ICE

The well known rock cycle describes the ways in which natural processes degrade and disperse geological materials, sorting their components and converting them into the raw materials of new, and often very different, rocks. At the earth's surface, meteorites are further from chemical equilibrium than most terrestrial rocks and therefore are more susceptible to destruction by weathering and dispersal of their components. They are also very rare among the overwhelming background of terrestrial materials. This is not true of the meteorites we find in Antarctica – they resist weathering for long periods of time and are often found in high concentrations on exposed patches of bluish ice at, or above, about 2000 m elevation. If a patch of blue ice contains a concentration of meteorites we call it a *meteorite stranding surface*. To a meteoriticist, the levels of concentration are almost unbelievable: as of December, 1999, Japanese, US and European field teams had searched only about 3500 km^2 of blue ice and recovered around 17 800 meteorite specimens.

Until recently, much emphasis had been put on the "treasure trove" aspect of the meteorite finds and their demonstrated value as scientific specimens, while very little attention had been given to the "treasure chests," i.e., the sites where they are found. These sites and the meteorites found on them are linked to the history of the ice sheet and to climate change. In addition to the many traditional

---

[1] Some sentences, paragraphs and photos in this chapter have been excerpted from a paper by Cassidy, Harvey, Schutt, Delisle and Yanai titled, "The Meteorite Collection Sites of Antarctica." This paper appeared in the journal *Meteoritics* in 1992. Excerpts appear without specific attribution but with permission of The Meteoritical Society.

areas in which meteorite research can make significant contributions to science, the history and dynamics of the ice sheet is another.

OPPORTUNITIES MISSED

Entire expeditions, as well as some individuals, have at times come close to discovering meteorite concentrations on the ice in Antarctica, but have not. In the following paragraphs I include some of these accounts.

Valter Schytt[2] and Charles Swithinbank

One of the earliest studies of blue ice fields has been reported by Valter Schytt in *Scientific Results of the Norwegian–British–Swedish Antarctic Expedition of 1949–52*. Valter's field party investigated some blue ice patches in Dronning Maud Land. They were interested in these occurrences as indicators of how the inland ice sheet reacts to possible changes in climate. They correctly identified these areas as sites at which deflation of the ice sheet was occurring. Measurements there indicated very low flow rates for the ice. Some of these sites were above 2000 m in elevation. These are all characteristics of meteorite stranding surfaces, and there is a good chance there were meteorites lying unrecognized on the ice.

Charles Swithinbank was a British member of the field party. Charles was to be active in antarctic research for many years, and the interested reader should know that he has published several very readable books about his experiences. Before 1988 he had become aware from our work that concentrations of meteorites could be found on some blue ice surfaces. During that field season he had occasion to visit a blue ice field adjacent to Mt. Howe, which, at 87° 22' S, is the southernmost rock outcrop in Antarctica. A member of his field party found a piece of iron lying on the ice, which Charles immediately recognized as a meteorite. We were camped at the Lewis Cliff Ice Tongue at that time, and a day or two after Charles' discovery, a twin otter landed nearby and taxied up to the outskirts of our field camp.

---

[2] Yes, it *does* appear to be pronounced that way.

Charles Swithinbank stepped out and presented the specimen to us. He also flew two of our party back to Mt. Howe to search for more specimens. John Schutt and Ralph Harvey scoured the ice patch over a seven-day period, but found only three more meteorites. However, of the four specimens recovered there, the iron is an ungrouped[3] ataxite while another is a eucrite, and a member of the rare and interesting HED achondrite group. By his discovery at Mt. Howe, I consider that Charles Swithinbank has compensated for his possible earlier oversights on the blue ice fields of Dronning Maud Land. I have not heard from Valter…

Scott's and Shackleton's South Pole expeditions

The Beardmore Glacier is a fearsome sight from the air, and even more awe-inspiring from the surface. It is highly dissected by crevasses and carries a heavy load of rock debris arranged in medial moraines parallel to its flow direction. Each medial moraine marks the upstream contribution of a feeder glacier tributary to the Beardmore. This tremendous glacier drains ice from a significant part of the ice plateau of East Antarctica, descending initially through a crevassed wilderness called the Wild Icefalls. This could be taken as an apt description of the feature, although it is named with reference to nearby Mt. Wild, which, in turn, was named after Frank Wild, a member of the British Antarctic Expedition of 1907–09.

The Beardmore carries ice through the Transantarctic Mountains at a tremendous rate, measured in places as high as meters per day. It flows down just east of a massive mountain called The Cloudmaker and debauches onto the Ross Ice Shelf to contribute its flow to the ice moving out to sea on both sides of Ross Island. In trying to reach the pole, the Scott and Shackleton expeditions used this glacier as a ramp to ascend through the Transantarctic Mountains to the ice plateau. It was on the Ross Ice Shelf just below the Beardmore that Scott and his two remaining companions, struggling

---

[3] "Ungrouped": a unique specimen. It may be the only sample we have of its parent body.

back from the pole barely alive, made their final camp and died in a storm, only 18 km from One Ton Depot – a cache of food and supplies that could have saved their lives.

As a footnote to history, their route up and back took them within 70 km of one of the densest concentrations of meteorites yet known in Antarctica – the Lewis Cliff stranding surface. They also approached to within 70 km of one of the most important vertebrate fossil sites in Antarctica – Coalsack Bluff. Neither of these scientifically important sites would be discovered until much later. It is unfortunate for Scott and his comrades that attaining the South Pole at that time was considered more important than making geological studies in Antarctica. If they had turned right at the head of the Beardmore they might have discovered one or both of these sites *and* returned safe and in good health.

B. M. Gunn and Guyon Warren

The principal objective of the Trans-Antarctic Expedition, 1955–1958, was to make an overland transit across Antarctica, from the Weddell Sea to McMurdo Sound, by way of the Geographic South Pole. A British party established Shackleton Base on the Filchner Ice Shelf of the Weddell Sea, where eight men wintered over during 1956. At the same time, a New Zealand party made a reconnaissance of possible base sites in McMurdo Sound that would be accessible to vehicles descending from the ice plateau of East Antarctica at the end of the traverse. In 1956, they established Scott Base at Pram Point on Ross Island, a short distance from the US base at McMurdo. During January and February of that year, they established a route up onto the ice plateau via the Ross Ice Shelf and Skelton Glacier. The route was convenient enough so that during the following field season they were able to establish food and fuel caches on the ice plateau by using tractor trains, aided by aircraft. This freed up four expedition members and two dog teams to work independently. During the austral summer of 1957–1958, B. M. Gunn and Guyon Warren were members of the New Zealand party. They, and two others, were able to spend the spring and summer of 1957–1958 engaged in a geological and topographical survey in the area north of the Skelton. They were designated as the Northern Survey Party.

Four men with two dogsleds crossed McMurdo Sound on the ice and ranged northward along the coastline to Cape Roberts, where they made a couple of side trips before ascending to the ice plateau via the Debenham and Miller Glaciers. Crossing the MacKay Glacier on a northward tack, they established a base camp near a small nunatak about 6 km north of Carapace Nunatak[4]. From this camp they made side trips to Carapace Nunatak, the southern tip of Allan Hills, Coombs Hills and Battlements Nunatak. On the trip to Battlements they very precisely skirted the western edge of the Allan Hills Main Icefield, upon which the ANSMET teams later discovered hundreds of meteorite specimens.

We came across their old campsite. In visiting sites of historical interest, where one's predecessors have been and have beaten the odds at the expense of great effort to return with scientific data, one can get an eerie sense of *presence*, particularly if one's imagination is given free rein. It is as if these people have left behind some essence of their having been there, and having striven, and some intangible trace of this has been impressed upon the earth or the rocks of the place and it will always be there. One expects sometimes to come face to face with them, or to hear conversations carried on the wind.

At the Northern Survey Party's base camp near Carapace Nunatak there was a more tangible presence – old socks blowing about on the ice. Old black socks, well-used, with holes at the heels and toes. A broken ski sticking up from a crack between two rocks into which it had been jammed in December of 1956 marked a cache left behind by departing explorers. It contained boxes of dog pemmican and trays of small glass vials containing what we suspected was cod liver oil. And old, discarded socks that had been plucked out of a box by the

---

[4] In Chapter 3, I describe our alarm at temporarily losing our field camp in a mirage at Carapace Nunatak.

incessant, insistent attentions forced upon them by the wind, which now blew them around in an icy hollow below the rock promontory on which the cache had been established. Old, used, holey, probably unwashed, discarded socks, and on each sock a small, oval-shaped label that read, "B.M. GUNN."

We took the socks back with us to camp, and afterward to McMurdo, packed away with our clothing, and after that, at the end of our field season, to Christchurch, New Zealand. In Christchurch we gave them to Hugh Logan, a scientist who later became head of the New Zealand Antarctic Programme, and who is a bit of a historian. He said, "Do you realize that Gunn is still alive, and active?" I hadn't even thought about that possibility. Hugh was delighted with the find, and proposed to wash them, and present them to Gunn at his first opportunity. The following year I checked with Hugh to learn what Gunn's reaction had been. After a little embarrassed hesitation, Hugh conceded that he hadn't given the socks to Gunn, but that he still planned to do so, at some propitious moment. Meanwhile, the socks were tucked away in a box in his garage, along with some other antarctic historical artifacts. The appropriate moment never came, however, and a few years later Hugh's wife made him throw the socks away because they were rotting. I had had the feeling that those socks somehow *were destined* to get back to B. M. Gunn. It was not to be; in the same way, I guess, that Gunn and Warren were destined not to find meteorites at Allan Hills.

James Collinson

In 1969, an Ohio State University expedition under David Elliot established a base camp near Coalsack Bluff, a highly productive vertebrate fossil site that David had discovered. Coalsack Bluff overlooks the Lewis Cliff Ice Tongue, 10 km away. An expedition member, Jim Collinson, actually visited the ice tongue. When he heard in 1985 that the ice tongue was home to hundreds of meteorite fragments he told me with a sense of wonder that he and his field assistant had spent about half a day walking around there and had not seen one meteorite!

Actually, he probably saw dozens, but he was looking for bone fragments and anything that looks like a meteorite is not a bone fragment.
I often wonder how many bone fragments *we* have looked at and not
seen because they do not look like meteorites!

Others

Other expeditions, criss-crossing Antarctica, have passed close to meteorite stranding surfaces without realizing their significance and
without finding meteorites. I have a theory that part of the reason for
this is that until recently, sledges were pulled by dogs, and dogs cannot pull anything over ice, because of a lack of traction. Sledge drivers
can be expected to detour around patches of ice. During long periods
of enforced inactivity during storms I constructed a mental picture
of sledge drivers trying to train their dogs to wear small crampons
on their paws, so they could navigate icy patches. The picture fell
apart, however, when I imagined the first dog wearing crampons sitting down and attempting to scratch behind his ear (or anywhere else).

While a number of explorers might have found meteorites in
Antarctica, but did not, French chemist J. L. Proust, actually suggested in 1805 that meteorites come from polar regions. This speculation came about when Proust was given a fragment of the Sena meteorite to analyze. The stone is an H4 chondrite that had fallen near
Sarinena, Spain, in 1773. Proust noted that the specimen contained
abundant small inclusions of nickel–iron alloy that were very prone
to rust upon exposure to water, and correctly deduced that stones
like it would not persist very long under ordinary conditions on the
earth's surface, where there is an abundance of water. A summary
of his report appeared in English in the *Journal of Natural Philosophy* (Nicholson's) in 1805. This reference, from which the following
quotation was taken, was pointed out to me by Derek Sears, of the
University of Arkansas.

> On considering the rapid alteration of these stones by moisture,
> for a fragment kept twelve hours under water was taken out
> covered with spots of rust, which distinguished the grains of alloy

from the sulphurous particles with which they were before confounded; – it is obvious, according to the author, that they cannot subsist in any of the habitable parts of the globe. But from the eternal cold of the polar regions, where water remains forever a solid mass, and iron cannot rust, he thinks we may reasonably look to these regions as the native place of such bodies. In this he insists there is nothing impossible, or even improbable. And why should those meteors, he demands, of which we know neither the origin, the combustibles that afford them aliment, the impulse by which they are moved, nor the nature of the lines they describe in their course, be less capable of tearing them from a part of the globe, than of forming them, contrary to all physical probability, from elements which the atmosphere can neither create nor hold in solution?

Meteors are now defined as the visible phenomena that accompany the fall of a meteorite. My reading of the quotation suggests that Proust had been told that meteors were the agents that form meteorites in the atmosphere. He argues that it is just as likely that meteors, which no one understood anyway, could tear these rocks loose from their sites of origin in the polar regions and transport them to habitable parts of the globe. At this point, Proust had the opportunity to suggest that someone go to the polar regions in search of more of these interesting rocks. He could have, but he did not.

BLUE ICE FIELDS: A HISTORY OF PROPOSALS

The geologists and geochemists who serve as investigators on the ANSMET project are primarily interested in meteorites. As gatherers of meteorite specimens, we are interested in why and how these wonderful concentrations of meteorites came to be. Their relation to the blue ice fields of Antarctica is overwhelmingly obvious. It is impossible to understand the cause(s) of these concentrations without also understanding the occurrence of the blue ice fields themselves. If there has been a disappointing aspect to the overall ANSMET experience, it is that the National Science Foundation has consistently rejected those parts of our research proposals in which we proposed to study the stranding surfaces. My understanding of fundamental research is that one's activities should be able to follow whatever direction seems indicated by current results. They made it clear to me, however, that even the simplest kinds of glaciological activities by ANSMET personnel would not be supported. I got the message, and included the following section in a proposal for five-year funding that I submitted in 1983.

> *Work not proposed*
> A number of reviewers of earlier ANSMET proposals had comments for and against the need for glaciological studies in connection with meteorite stranding surfaces. Some pointed out that ice dynamics studies could help us locate new meteorite concentration sites because we would then better understand the mechanisms involved. Another reviewer criticized the proposed limited interaction with glaciologists. One reviewer, however, felt that signed letters from glaciologists should accompany my assertion that ANSMET meteorite collections also have great potential in ice sheet studies. This same reviewer felt that ANSMET should perhaps be terminated, existing meteorites thoroughly analysed, the results published, and then the need for further glaciological work established. Reviewers' suggestions ranged from terminating the ANSMET program (mentioned above) to incorporating glaciologists into the field teams and/or funding a postdoctoral position at the University of Pittsburgh to be occupied by a glaciologist.

The section quoted above hints at how painful it can be to read reviewers' comments. Forgetting pain and some of the negative reviews, however, a recurring theme among some was agreement on the need for a glaciological component to the ANSMET project. I already knew what the official view was at the Office of Polar Programs, and therefore tried to forestall the same kinds of comments on this proposal with the following paragraph.

> The number of reviewers who offered comments along these lines (i.e., suggesting glaciological studies) was impressive, and I felt they could not be ignored. After careful consideration, however, I decided (1) that continuation of the ANSMET project could be justified based only on our meteoritical results to date, and (2) that it would be inappropriate for me to propose a significant component of glaciological study as part of the ANSMET project. I have tried to indicate in the present proposal that glaciological studies could be a logical complement to our ANSMET activities. I would expect that such studies would be proposed independently by others, and I would expect to stay in close contact with such investigators in order to benefit as much as possible from their findings. If, however, independent glaciological studies are not proposed, or are not funded for some reason, I would then consider requests from individual glaciologists to be part of the ANSMET field team, and would seek NSF approval of such a plan.

This proposal was ultimately supported. One reviewer of this proposal referred specifically to the quoted comments of those earlier reviewers who had recommended against glaciological studies and the one reviewer who had proposed terminating the ANSMET project. I particularly appreciated this reviewer's point of view.

> The comments from glaciologists are somewhat of a surprise. The ANSMET project can certainly be justified without glaciological applications but I would have expected a more enthusiastic response. (I am reminded of the substantial number of geologists of 20 to 50 years ago who dismissed any suggestion of the importance of terrestrial or lunar impact features. Can these be their intellectual descendants?)

Even though the Office of Polar Programs was unwilling to support any incursions into the glaciology of stranding surfaces, I had a secret weapon – they could not stop us from *thinking*, and this often led to some small projects to which the funding agency turned a blind eye, as long as they had not been officially proposed.

Contrary to my hopes, there has not been a surge of interest in stranding surfaces among the members of the glaciological community, with one outstanding exception: Ian Whillans, at the Byrd Polar Research Center of The Ohio State University.

Ian was a glaciologist. He was intrigued by meteorite stranding surfaces as soon as he heard of them, and we had a very amicable collaboration. He often spoke of plans to propose glaciological research directed toward understanding stranding surfaces, but complained that his time was currently committed to major ice stream studies in West Antarctica. Ian died in 2000, at the height of his career, without making the contribution to stranding surface studies that we had hoped for. A few other workers should be mentioned.

Gunter Faure is a radiochemist with a bias toward field work that involves the ice sheet. He has carried out some ice movement and ablation studies at the Lewis Cliff and Elephant Moraine stranding surfaces, and at the latter site has studied in detail the moraine rocks that make up the elongated "trunk" of the elephant. This study was especially clever, I thought, because he was trying to infer relationships among inaccessible rock formations under the ice by studying the debris from these formations that could be found as morainal deposits. I was bemused by the fact that this is exactly what we do with meteorites, whose parent bodies are either inaccessible or destroyed.

Ice movement studies have been made at the Allan Hills Main Icefield by John Annexstad, then at NASA's Johnson Space Center and Georg Delisle at the Bundesanstalt für Geowissenschaften und Rohstoffe, at Hannover, Germany, so we have some information on ice flow at that site.

During the 1993–94 field season we hosted Keith Echelmeyer, then at the University of Alaska, who led a radio echo-sounding study of the bedrock topography underlying the Lewis Cliff Ice Tongue stranding surface. His results contributed in a significant way to our understanding of the distribution of meteorite specimens on the ice surface above. He traced two separate, parallel bedrock channels below the ice tongue. Most of the meteorites we found on the ice surface were located over the western trough. Very few were found on the ice over the eastern trough. We surmised that the western channel followed the course of ice coming off the ice plateau of East Antarctica. This flow must have a very long upstream collecting area, and the residual concentration of meteorites on the surface reflected this large source area. The eastern channel, however, probably drained only the Walcott Névé which, because of its local origin and small areal extent, had collected few meteorites.

TERMS AND CONVENTIONS

Georg Delisle has suggested the term *meteorite trap,* instead of *meteorite stranding surface,* to describe the patches of ice on which concentrations of meteorites are found. I use meteorite trap in a general sense to include the entire volume of ice whose meteorites eventually will reach a stranding surface, plus the stranding surface itself, plus any physical features such as barriers or bedrock channels that contribute to the trapping process. Meteorite stranding surface continues to be used to describe one component of a meteorite trap.

Stranding surfaces very often display one or more steps downward along the direction of ice flow. In early papers I used the term "monocline" to describe such a feature, but this wrongly implies folding, in a structural sense. Another term in use is *escarpment.* This implies a steep slope, and most of these slopes are relatively gentle (see, for example, Figure 10.1). *Ice ramp* is the preferred term, but in the absence of a suitable synonym, escarpment is not completely abandoned.

A meteorite may fall onto the ice intact, it may break apart in the atmosphere and produce a shower, and/or it may fragment on impact or fragment later due to weathering. A shower of meteorites produces a *strewnfield.* Here I use the term *scatter field* to indicate a scattering of meteorite specimens on the ice that results from any of

FIGURE 10.1 An ice ramp at the Allan Hills Main Icefield meteorite stranding surface, photographed from the basin below the ramp. Notice the characteristic windblown snow patches that often partly mask the ice surface at these meteorite stranding sites. (Photo by L.A. Rancitelli.)

the above causes, including a shower. This is useful because often we cannot identify the members of a shower on antarctic ice, or differentiate them from fragments produced by processes other than breakup in the atmosphere. The question of *pairing* is a constant problem on meteorite stranding surfaces because the density of occurrence of meteorites can be very high, and scatterfields can overlap.

WHERE ARE THE BLUE ICE FIELDS?

Antarctica can be thought of as having three major divisions: the Antarctic Peninsula, East Antarctica and West Antarctica (see Figure 1.1). The Antarctic Peninsula extends north from about 73° south latitude and has a relatively mild climate, compared to the rest of Antarctica, with comparatively high levels of precipitation. It is often referred to as "the banana belt," in recognition of its relatively milder climate. East and West Antarctica make up the major part of the continent, with most of East Antarctica located in the Eastern

Hemisphere. East Antarctica is a vast polar desert, largely covered by a thick lens of ice mantled by a thin layer of snow; it is the only part of Antarctica in which meteorites have been found in any numbers. The highest point on the ice plateau of East Antarctica is near the center of the continent (not the South Pole) at a site called Argus Dome. The elevation here is about 4000 m. To a first approximation, the ice surface slopes downward from this point toward the edges of the continent. Frigid air slides downhill over the surface as density currents, accelerating with distance and producing the so-called *katabatic winds* near the edge of the continent.

By far, the largest surface feature interrupting the vast expanse of snow-covered ice on the main part of the continent is the Transantarctic Mountain Range, marking the boundary between East and West Antarctica. Moving from the Transantarctics into the ice plateau of East Antarctica, the mountains rapidly degrade into lonely peaks, emerging from the ice as points or ridges called *nunataks* that act as largely ineffectual barriers to the encroaching ice. They do, however, presage a major contortion of the ice sheet as it is forced to funnel into narrow passes through the mountain range. Within the zone between the nunataks and the beginnings of these outlet glaciers, at an elevation around 2000 m, are scattered patches of blue ice that are continuously swept clear of snow (Figure 10.1) by katabatic winds. Some of these ice patches have concentrations of meteorites on them (Figure 10.2). Meteorite concentrations not associated with the Transantarctics occur in Queen Maud Land, which is also in East Antarctica. The Queen Maud Land concentrations also are associated with large barriers to ice flow and katabatic winds (Figure 10. 3). Aspects of three of the major meteorite stranding surfaces are described below.

THE METEORITE STRANDING SURFACE AT THE ALLAN HILLS MAIN ICEFIELD

The Allan Hills Main Icefield can be seen in the Frontispiece. The katabatic winds not only keep the ice clear of snow but also

FIGURE 10.2 Field curation of meteorites at the Allan Hills Main Icefield. (Photo by R. Walker.)

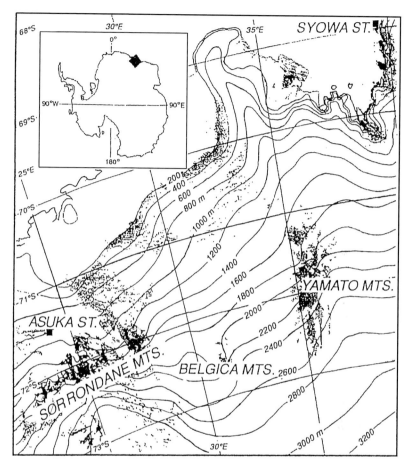

FIGURE 10.3 Regional map, including the Yamato, Belgica and
Sør Rondane Mountains, in Queen Maud Land. Filled areas are sites
where rock crops out above the ice. Note the locations of the Japanese
stations Asuka and Syowa. Inset shows the location and orientation of
the map relative to the antarctic continent. The surface of the ice sheet
descends from 3200 m south of the Yamato Mountains to 200 m where
the underlying rock is at sea level. The direction of ice movement at any
point can be estimated by drawing an arrow downslope perpendicular to
the nearest topographic line.

concentrate the smaller meteorite fragments in a downwind direction at this 75 km^2 patch of blue ice lying just west of Allan Hills in South Victoria Land. In the 1984–85 field season John Schutt arranged a field test of the windblown effect, by setting up a "rock race." He established two rows, or "teams," of terrestrial rocks in lines perpendicular to the prevailing wind direction. The rocks in one row were placed on the surface and those in the other row were frozen entirely into the ice. By the following austral summer, all except the three largest rocks in the subice team had emerged completely from the ice and these three, ranging up to 804 g, were almost completely free, indicating that at least 4 cm of surface ice had ablated away. The eight smallest rocks in this team, ranging from 1 g to 18 g, had all moved from their initial sites and five of them no longer could be found. The nine smallest rocks in the surface team had moved and three of them no longer could be found. To date, in both teams all rocks up to 70 g have moved over the surface. From this experiment, therefore, one could expect that all meteorites weighing less than about 100 g may have migrated downwind on any stranding surface.

Flow on the ice sheet follows the direction of descending surface contours, reflecting the fact that gravity is its motive force. Only in areas where ice flow is inhibited by a barrier is there believed to be any lateral force exerted by flow. Where the ice cover thins, bedrock irregularities can influence flow direction, as long as the ice is not immovably frozen to the underlying surface. This influence of bedrock on ice flow may be the case at the Allan Hills Main Icefield, where radio depth-sounding measurements suggest a channel running northward along the upstream edge of the icefield, across the direction of arriving ice. Movement studies of the surface ice indicate that much of the ice flowing toward Allan Hills is diverted northward parallel to the subice channel, and only a smaller fraction of the ice continues onto the icefield and over the barrier toward Allan Hills. Between the barrier and Allan Hills is another bedrock channel trending northward, and ice that has escaped the ablation process on the Main Icefield presumably then could follow this channel northward.

Where meteorite stranding surfaces exist but there is no nunatak cropping out above the ice as evidence of a barrier to flow we can infer the presence of a subice barrier. This is indicated by ice ramping downward, as if spilling over the lip of a dam. See Figure 10.1, which shows the ice ramp at the Allan Hills Main Icefield. Ice flows down this ramp and into an ablation basin. Meteorites are exposed on the ice surface before the barrier, at the barrier and on the ramp below the barrier, but are found in the greatest numbers in the ablation basin below the ramp. The fact that meteorite concentrations are high in the ablation basin suggests that ice is constantly being supplied to the basin, and ice is flowing out of the basin at a very slow rate, or not at all.

This conclusion reflects only the present conditions. There is no guarantee that ice movement velocities and ablation rates have always been the same, but it seems reasonable that these conditions are typical of the present interglacial stage. There is some evidence that ice levels rose 50–100 m higher against the western side of Allan Hills. This suggests that at an earlier time the ice sheet at Allan Hills was thicker.

Exactly when that was, is open to discussion. Ian Whillans points out that climate warming produces ice sheet thickening because of increased precipitation in inland areas. As that thicker ice moves downstream, it can lift a stranding surface, along with its accumulated meteorites, over the barrier. Later, as the warmer climate persists, the ice warms up, becomes less viscous and flows faster. This causes thinning of the ice sheet and the stranding surface can be reestablished, but the earlier collection of stranded meteorites is gone. What a loss! If only something could be done to correct the profligacy of Nature!

Accumulation of meteorites begins again, of course, and continues through any subsequent climate chilling until the ice sheet thickens again during the next interglacial.

THE YAMATO MOUNTAINS METEORITE
STRANDING SURFACE

The Yamato Mountains is listed on most maps as the Queen Fabiola Mountains, but I use here the name indicated on the Japanese maps. To reach the Yamato Mountains, field parties start from the principal Japanese base, Syowa Station, located on the coast south of the Indian Ocean. Traveling overland, they follow a somewhat circuitous route to avoid the drainage basin of the Shirase Glacier, with its associated crevasses; finally reaching the Yamato Mountains at a location about 300 km south-southwest of Syowa Station. The mountains protrude above the ice at seven major locations along a 50 km north–south trending arc. From south to north, these points are designated Massifs A–G. The first antarctic meteorite concentration was discovered by Japanese glaciologists on the icefield at the Yamato Mountains.

The Yamato Mountains are surrounded by a blue ice field whose area is about 4000 km^2. This is by far the largest blue ice field on the Polar Plateau, and by the December 1998 field season, the Japanese field teams had reported almost 6000 meteorites and meteorite fragments found on the estimated 1500–2000 km^2 part of this icefield that they had searched. The meteorite stranding surface is elongated in the north–south direction and consists of several large blue ice fields and many smaller patches. The icefield is gently to steeply sloping and receives ice flowing from the Polar Plateau toward the continental margin. The ice surface descends from south to north in a series of ramps aligned southwest–northeast, from an elevation of 2700 m to 1600 m. The actual exposed area of the Yamato Mountains is dwarfed by the expanse of the icefield; particularly in the upstream direction to the south, where the mountains are almost engulfed and only small nunataks crop out.

The meteorite stranding surface contains topographic depressions, dark-colored streaks (dirt bands), wind scoops, ice mounds and moraines; also linear and curved crevasses that can be used to interpret ice flow directions. Crevasses occur most commonly around nunataks and on steep slopes, and less commonly on gentle slopes. They are seen bridged with snow on the blue ice surface and also are detected in areas where snowfields tend to disguise them. They are typically from a few meters to 10 m wide and are an ever-present danger.

The southernmost (i.e., farthest upstream) exposure of the Yamato Mountains is the Minami Yamato Nunataks group. The ice there is horizontal to gently sloping. Dust bands are common, generally as linear features a few centimeters to a few tens of centimeters thick, although occasionally they occur as patches and in contorted patterns. Aside from rare dirt bands clearly associated with nearby moraines and containing coarse-grained rock fragments with fines, these dust bands are volcanic ash containing glass shards and mineral fragments in the size range 20–40 μm.

To reach the Yamato Mountains, ice flows north-northeastward about 550 km from a topographic divide on the ice plateau. At the Yamato Mountains barrier it flows around and between the massifs in a generally northwest direction. The mountains must have an extremely wide subsurface expression because the blue ice field over them is so large. Measured ablation rates are 2–7 cm/year, depending on the location. The average of all measurements is about 5 cm/year. The loss of ice is almost exactly balanced by the rate at which replacement ice is welling up.

THE LEWIS CLIFF METEORITE TRAP

In this region, the principal release for ice coming down off the polar plateau is the Beardmore Glacier (Figure 10.4), a chaotically crevassed river of ice flowing down the eastern side of the north-northeast to south-southwest trending Queen Alexandra Range. At its head, the Beardmore collects ice converging from the south and southwest. Ice spills in off the polar plateau, but at some point the collection pattern is determined on the west by the southern extremity of the Queen Alexandra Range, cutting into the advancing ice sheet like the blunt prow of a supertanker. The alternative outlet for the ice thus diverted is the Law Glacier, 70 km distant to the north. Here, ice trying to turn the corner into the Law, which is flowing in a relatively stagnant

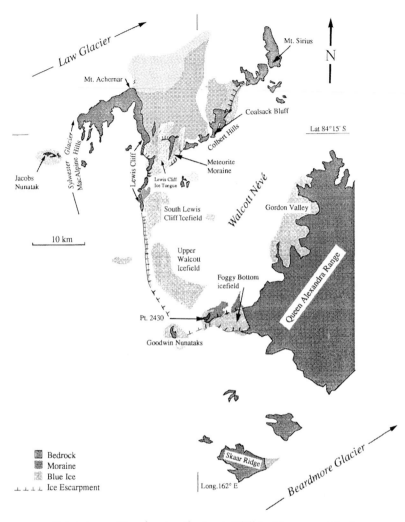

FIGURE 10.4 Map showing the Lewis Cliff Ice Tongue and the features surrounding it. Note that the ice tongue extends into a massive moraine, which envelopes it on three sides. Note the nearby location of the Coalsack Bluff *Lystrosaurus* fossil locality, one of the truly significant antarctic vertebrate fossil finds, with important implications to plate tectonics. Waiting to be found only 10 km from Coalsack Bluff was the Lewis Cliff Ice Tongue, with its rich harvest of meteorites. Only 70 km distant are the routes the Shackleton and Scott parties followed up the Beardmore to reach the ice plateau.

manner, is suffering ablation – the result is the MacAlpine Hills stranding surface.

Between these two major outlets is a zone of relatively stagnant flow that is being drained by neither the Beardmore nor the Law, and the ice diverted from the Beardmore adds itself to this mass. Topographically, in this zone the ice surface lies at greater than 2400 m elevation, allowing it to flow over an apparent subice barrier or cirque rim into the southern edge of the Walcott Névé. The barrier appears above the ice at three points: at an unnamed nunatak (Map Point 2430), at Goodwin Nunataks, and at Lewis Cliff. This is the only external source of ice flowing into the Walcott Névé; all others are local sources from the flanks of the Queen Alexandra Range and from snowfall directly onto the Névé surface. The Névé appears to be a fairly stagnant mass of ice with some drainage through a narrow pass east of Mt. Sirius into the Bowden Névé to the north, and with minor amounts escaping northeastward through the Colbert Hills.

Possible meteorite collecting surfaces associated with the Walcott Névé are limited to some blue ice fields along the edge of the Queen Alexandra Range, where we encountered only two specimens, and some blue ice fields above and below the submerged parts of the Colbert Hills, where we succeeded in finding only one meteorite, below the ice ramp. This supports the view that the main body of the Walcott Névé is not a productive area because it gathers snow only from local sources. By contrast, all the known meteorite stranding surfaces probably derive ice from the ice plateau of East Antarctica.

According to this interpretation, ice from the Plateau, some of it very ancient for having traveled a long way, encounters the main mass of relatively stagnant snow-covered ice filling the Walcott Névé, is shunted northward and flows in a stream below and parallel to the Lewis Cliff escarpment. Eventually it reaches the Lewis Cliff Ice Tongue. This is the end of a long migration pathway. Here it stops until it wastes away by ablation. Looming over the Lewis Cliff Ice Tongue is Mt. Achernar. It seems particularly appropriate that the translation of this Arabic name means *The End of the River*. Both rocks and meteorites are segregated toward the western side of the ice tongue, perhaps indicating that ice on the eastern side has the less productive main body of the Walcott Névé as its source.

Shown in Figure 10.5 are areas in which we have recovered more than 3700 meteorite specimens (as of the December 1998 field season); almost 2000 of them on the Lewis Cliff Ice Tongue itself. Figure 10.5 is a map of finds at the Lewis Cliff Ice Tongue where, because of the small scale of the map and the large number of recoveries, dots are sitting on dots.

Between the South Lewis Cliff Icefield and Coalsack Bluff, in an area where meteorites were sparsely distributed, were dark objects that excited our imaginations because from a distance they looked like large meteorites. They turned out to be empty JATO pods[5] jettisoned many years earlier by LC-130 cargo planes after the punishing lift-off run from the 1960s' camp at Coalsack Bluff. It is truly difficult, these days, to "go where no man has gone before!" It is thought-provoking also to reflect on the fact that large cargo planes may have been taxiing about among meteorites.

THE LEWIS CLIFF ICE TONGUE METEORITE STRANDING SURFACE

The Lewis Cliff meteorite stranding surface is a classic example of the catchment area of an absolute trap for meteorites. At 2000 m altitude, a tongue of ice about 2 km wide and 9 km long, its surface slopes downhill toward the north, with a steeper ramp down occurring

---

[5] JATO (Jet-Assisted Take-Off) pods are devices that can be mounted on the outside fuselage of an LC-130, two on each side. When mounted, they are tilted upward to give additional, sudden thrust to help lift the plane off the surface. The horizontal component of their thrust helps it to reach flight speed quickly. Once they have burned completely they can be dropped off. Beautiful in concept, they do not always perform flawlessly. An LC-130 attempting to take off from a site named Dome C had one come loose during the taxi run and shoot off one of the engines. Fragments of the propellor crashed through the cabin, and the pilot had some difficulty in aborting the takeoff because the other three JATO pods were still burning, but the thrust was asymmetric because one pod had been lost. The ultimate track of the plane on the snow looked like a giant "J", with the plane sitting at the end of the hook-shaped track.

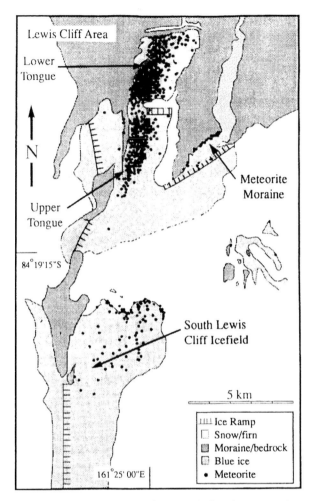

FIGURE 10.5 Map of meteorite distribution on the Lewis Cliff Ice Tongue, Meteorite Moraine and South Lewis Cliff Icefield. Note the clustering of meteorites toward the western side of the ice tongue. Radio echo-sounding measurements over the ice tongue indicate two parallel channels in the subice rock surface. The eastern one presumably carries ice coming off the Walcott Névé, which has a relatively small collecting area and therefore is not very productive of meteorites. The western one presumably carries ice from the ice plateau, which has a large collecting area for meteorites. The source of the relatively dense concentration of specimens along the edge of Meteorite Moraine is probably the Walcott Névé. Most of these are small fragments, probably from only a few falls.

FIGURE 10.6 Evaporate band decorating a crack in the ice. The light tan crystals are in large part sodium bicarbonate (nahcolite; better known as baking soda). (Photo by W. Cassidy.)

partway along its length. Thus we speak of an "upper tongue" and a "lower tongue." The Lewis Cliff Ice Tongue (LCIT) is bounded on the east and west by morainal and bedrock ridges. At its northern (downstream) end, the ice tongue is separated from the Law glacier by a vast, complex moraine (see Figure 10.8).

Many small, frozen-over, or completely frozen, meltwater ponds are scattered among the moraine ridges near the margins of the Lewis Cliff Ice Tongue. Evaporites are prominently visible around the ponds, but also as apparent efflorescences among the rocky debris covering the ridges (Figures 10.6 and 10.7). Evaporite bands on the ice surface can be found in places associated with cracks near to and parallel to the edge of the ice tongue. We do not understand their origin, but all of these features probably indicate an ablative environment and a very briny water source.

From my experience with glacial moraines in the northern hemisphere I had a mental image (if I thought of it at all) of linear

FIGURE 10.7 The large whitish masses on the morainal ridge are mixtures of a number of evaporite-type crystal species, including nahcolite and borax. (Photo by R. Korotev.)

piles of rock that had been carried along by a glacier and deposited along the glacial front, leaving arc-shaped rows of rocks behind as it retreated in stages (terminal moraines), or along lines parallel to flow lines (medial moraines), or along the meandering courses of subglacial streams (eskers). I didn't see any such features in Antarctica, although they may be present in the Dry Valleys. The moraines we found are described as ice core moraines. These are ridges of ice that are mantled by a veneer of glacially worked rocks. They look like a pile of rocks, but when one assumes it is a pile of rocks and tries to climb up on it, the rocks tend to slide, exposing bare ice underneath and sending one sliding back down. This is a fair description of the moraine ridges seen in Figure 10.8. They are ridges of ice with a thin covering of loose rocks.

I began to get an insight into the origin of these ice-core moraines when we came across one near the western edge of the lower tongue that had been sectioned by wind erosion (Figure 10.9).

FIGURE 10.8 Aerial view looking toward the south over the Lewis Cliff meteorite stranding surface, and the array of morainal ridges blocking its flow north, toward the bottom of the picture. The pattern of morainal ridges is outlined by snow deposits in gullies between the ridges. Linear white streaks extending north–south across the moraine are drift lines of blown snow deposited on the moraine. These are parallel to the prevailing wind direction down the ice tongue. Goodwin Nunataks can be seen near the horizon, just left of center. (Photo by W. Cassidy.)

The cross-section of the ice ridge of Figure 10.9 apparently has been exposed by wind erosion; the ridge extended originally through the position from which the photo was taken. The center of this feature is occupied by an almost vertical debris zone of glacially worked rock and unsorted rock debris, frozen in the ice. The slopes of this ice-core moraine are mantled by rocks derived from downward erosion of the central rock debris zone. The feature apparently stands out as a ridge because as initial ablation of the surrounding ice occurred, the rocks along the central debris zone were increasingly exposed and tended to fall out of place laterally. This detritus largely shielded the underlying ice from further ablation, creating a small ridge. As surrounding ice continued to be lost to ablation, the debris cover expanded laterally by gravity, protecting a wider and wider band on both

FIGURE 10.9 An expedition member standing on a frozen meltwater pond next to a vertical section through an ice-core moraine. This cross-section has been exposed by wind erosion. (Photo by M. Wieczorek.)

sides of the ridge, and allowing the ridge to grow ever higher relative to the surrounding, still ablating, ice.

Structures similar to this have been called Thule-Baffin Moraines, after type localities near the margin of the Greenland ice sheet and on Baffin Island. The usual interpretation of a feature such as the rock debris zone of Figure 10.9 is that it originated as a fault plane in the ice, in which an overriding slab carried up rocks from the bed of the glacier in a shearing process. In an interesting paper on this type of moraine, J. Weertman, at Northwestern University, Illinois, has pointed out that basal melting and refreezing would be necessary in order to embed the rocks in ice before they can be transported to the surface. Examination of the ice in which the rocks are embedded shows it to be almost bubble free, which does indicate that it has been melted, degassed and refrozen.

While ice is known to shear, this particular structure may not result from that process. Two bands of bubble-free, debris-bordered ice flank the central debris layer, one on either side. They are separated from the debris layer by apparently unmelted ice. If this were material carried up by shearing, one might not expect both sides of the shear zone to be symmetrical in this manner. An alternative explanation for the rock dyke therefore is that it is a very tight fold in which both outlying bubble-free layers are in fact the same feature, which was originally horizontal above the debris layer but now occurs on both sides of the fold. This is the type of interpretation a structural geologist might suggest, and whether it is applicable to glacial dynamics remains to be seen. But ice is rock being held near its melting temperature, so structures similar to those found in metamorphic rocks would not be unbelievable.

If the high point on this ice core moraine represents some original surface in existence before ablation occurred, then we can be sure from the level of the surrounding surface that 12 m of ice have been removed by subsequent ablation. This is a minimum estimate, however, because the rocks scattered over the surface of the moraine came originally from its rock debris core, suggesting that earlier it stood higher.

Whatever the origin of this feature, I assume that each moraine ridge in Figure 10.8 is similar to this one, and that ice-core moraines all have a central plane of rock debris. If they are formed by folding, instead of faulting, the massive, complex moraine at the end of the Lewis Cliff Ice Tongue is a series of wrinkle ridges in the ice.

To me, a noteworthy feature of the Lewis Cliff Ice Tongue is that this place located only about 640 km from the south pole, at an altitude of 2000 m, had liquid water in some of the frozen-over meltwater ponds, and growing in one of these ponds were algae. These plants, discovered by Ralph Harvey, were seaweed-like growths up to 15 inches (c. 38 cm) long, in perpetually motionless water, anchored in the bottom and stretching upward toward the light that was filtering through the ice. After discovering them, we mentioned them in our next radio communication with McMurdo and received permission to collect one example. Upon our return it was deposited in the collections of the Biolab at McMurdo, but I received the impression that it was no big deal.

The preceding discussion of ice-core moraines, evaporite deposits and frozen ponds has, in a way, been a side issue involving characteristics intrinsic to a single stranding surface. A more general question involves the origin of the Lewis Cliff Ice Tongue and other stranding surfaces as collectors of meteorites.

SETTING THE STAGE

Antarctica has not always been shrouded in ice. We know this from the fossil record of plant life that must have grown in a warmer climate, and the animal species that wandered its surface. None of these could have survived there today. At some point in the past, the entire continent grew colder and colder, killing off all of these warmth-loving life forms and, becoming colder still, was transformed into a polar desert in which snow fell sparsely but did not melt. The snowpacks on the highest peaks grew thicker while the snow on the lowlands accumulated apace. Mountain glaciers appeared and drained ice off the high points as it continued to accumulate in the lower levels. The growing weight of ice bore down on the land surface and the rock formations began to yield, by sinking deeper into the crust, until the rocky surface of the continent was largely at or below the level of the surrounding sea, and the mountains that earlier had shed ice by glacial flow were now largely buried in it. An all-encompassing ice sheet coated the surface and, as it grew thicker, began to flow outward from the center toward the seas under the influence of its own weight. This was, and is, the perpetual winter of the antarctic continent. This wintry surface is receiving meteorites.

Meteorites fall onto Antarctica at the same low but constant rate as they fall elsewhere. Unless it falls directly onto a stranding surface, the terrestrial history of an antarctic meteorite includes a period of time embedded in ice[6]. Whether a meteorite becomes part of a concentration in Antarctica apparently depends on its trajectory across the continent, carried by ice from the point where it fell to the point where it is ultimately freed of the ice surrounding it. Most are released from the ice when a floating iceberg melts, and it falls into sea floor sediments as a dropstone. A relative few of all that have fallen on the entire continent are exposed on a stranding surface when a mass of stagnant ice experiences negative mass balance and evaporates away, leaving meteorites behind as a sort of residual deposit.

Because of changes in worldwide climate, the antarctic ice sheet probably has been at times thicker, and at other times thinner, than it is today. As the antarctic ice sheet thickens and thins, meteorite stranding surfaces can form in different places. After accumulating meteorites for some length of time, a stranding surface either will be flushed out by ice thickening, or deposited on a mountainside where weathering erases the accumulated meteorites.

Meteorites traditionally have been sought as messengers from space. But in Antarctica meteorites may be messengers from the ice

---

[6] Meteorites can fall onto rocky surfaces, of course, but only the irons seem to be preserved for any length of time on such surfaces. We have rarely found stony meteorites on rock outcrops, and the proportions of such stones to irons recovered on rocks in Antarctica is very low. Stones apparently must be in or on the ice to be preserved.

sheet, because they are concentrated under special conditions having to do with ice flow and micro- and macroclimatic conditions. Their terrestrial ages can be measured, and they may carry a history from their terrestrial sojourns. Thus, we who use meteorites to try to gain insights concerning the primordial cloud, the solar nebula, the origin and evolution of planets and planetoids and the environment of the space between the planets now have another field in which these invaluable objects may be useful.

CHARACTERISTICS OF METEORITE STRANDING SURFACES

While all of the meteorite stranding surfaces discovered to date share some common characteristics, many sites exhibit additional features that are not common to all. Also, differing surroundings, weather, and history have given each stranding surface its own flavor, which in the context of the meteorites collected, offers insight into the processes of meteorite stranding surface formation and, possibly, vignettes of the history of the ice sheet. In preceding sections I have described what is known about the occurrence of antarctic meteorites at three important stranding surfaces. Study of these and other stranding surfaces has suggested a few generalizations.

- To be recognized, a meteorite stranding surface must have bare ice. Where bare ice is present the surface is being removed by ablation, so meteorite stranding surfaces are characterized by active ablation. Where it has been measured, ablation at meteorite stranding surfaces averages 5–6 cm/year.
- Since meteorite falls are so rare, extremely large volumes of ice must be processed through the ablation surface to build up a concentration, and this requires a very long period of time. Hence, if a meteorite stranding surface has many meteorite groups represented on it, it must have been in existence for a long time.
- Ice on which meteorites are concentrated must not have a downstream outlet, or else must be flowing out at a slower rate than new ice is flowing in. To retain a concentration of meteorites on

a stranding surface, the supply of ice must at least equal the loss by ablation and flow; if not, the underlying rock will be exposed and all the meteorites except the irons will soon disappear. Field observations suggest that ice always is moving into the ablation zone where a concentration of meteorites is building.

- The numbers of meteorites in any collection from a single stranding surface decrease exponentially as their terrestrial ages increase. This observation is discussed later, in some detail.

METEORITE CONCENTRATION MECHANISMS ON STRANDING SURFACES

- Direct infall.

Meteorites fall onto deflation surfaces just as frequently as they do onto snow accumulation surfaces. Therefore some fraction of any concentration found on a stranding surface results from direct infall. Estimates of observed falls in temperate zones suggest 6 falls/km^2/million years (Myr), resulting in an estimated 60 specimens/km^2/Myr after fragmentation in the air, or on impact. We can find smaller objects on the ice in Antarctica than elsewhere, because they are not hidden among grass, leaves or rocks. If these are fragments, they are accounted for among the estimated 10 fragments per fall. It is probable, however, that some of them are small individual falls that normally would not be found elsewhere. One might guess that this would raise the estimate by another factor of two, so that we might guess that direct falls onto a stranding surface in Antarctica could amount to 120 individuals/km^2/Myr.

- Ablation discovery.

Ice-wasting occurs by ablation. We find meteorite concentrations on surfaces of deflation, where active ablation is occurring. Because meteorites are such rare objects, if a body of ice is losing mass by ablation at its surface it will produce a concentration of meteorites only after large volumes have passed through the ice/atmosphere interface.

- The passage of time.

  Long periods of time are necessary before an appreciable accumulation of meteorites can be produced, even with additive contributions from direct infall. A meteorite fall is still a rare occurrence.

- Moderation of the weathering process by low temperature.

  Even more than most terrestrial rocks, meteorites are out of chemical equilibrium at the earth's surface, and stony meteorites weather very quickly. Chemical reaction in the presence of liquid water is the principal culprit. The intense cold of Antarctica, however, is an ally in ensuring that the meteorites see very little liquid water. Weathering still occurs, but at a much slower rate, so specimens survive longer. The lower the temperature of the environment on the stranding surface, therefore, the slower will be the chemical reactions that destroy meteorites, and the more favorable will be conditions for preservation and accumulation.

- Favorable flow rates.

  Meteorites accumulating on a stranding surface will be carried away if the ice is flowing away, therefore for a concentration to grow, the flow rate of ice into the area must be greater than any flow rate out of it. The concentration density will be sensitive to the size of the following ratio:

  flow rate in/flow rate out

- Wind.

  The wind is capable of moving small pebbles across the ice quite rapidly, and carrying them to the rougher firn surfaces at the downwind edge of stranding surfaces where their further progress is stopped. While the wind can, and does, rearrange the local concentration of small meteorites on a stranding surface, in general it does not change the total number of meteorites. A strong wind can drive small meteorites some distance across the firn at the downwind edge of the icefield and a later, milder wind could cover them with snow, but another strong wind is sure to uncover them again. This would not change the number of meteorites on the stranding surface, but it could change our estimate of that number and our opportunity to collect them, since at any given time some of them may be covered.

Another effect of wind, of course, is that it enhances the ablation process, and wasting of the ice favors a more rapid accumulation of meteorites on the surface.

To summarize this section, the factors we are aware of that conspire to produce accumulations of meteorites on the ice are ablation, the effect of wind in promoting ablation, direct infall, the passage of time, and cold temperatures. Weathering does nothing to increase the total mass of the concentration, but it can increase the numbers of fragments. The only factor we are aware of that acts to reduce concentrations is ice flow out of the stranding surface.

METEORITE STRANDING SURFACES AND
ICE SHEET DYNAMICS

Observations had suggested where to look for meteorite stranding surfaces, and also the characteristics we should expect them to have. This is outlined in the two sections above. Certain characteristics of the antarctic ice sheet also were pretty well known. Ian Whillans thought it would be a good idea at this point to try to integrate the stranding surface into the lore of the ice sheet. His idea was to propose a unified model for the entire meteorite/ice sheet interaction, starting with fall of meteorites onto the ice sheet and ending with their deposition on a stranding surface. The model would be consistent with what was known about ice sheet flow and what we had inferred about stranding surface processes. He warned that we should expect our ideas to be attacked and possibly discredited, but was confident that this type of dialogue would bring us ever closer to a full understanding of the stranding surface/ice sheet interaction.

I agreed that it would be a worthwhile effort, and we wrote a paper that appeared in *Science*, in 1983, titled, "Catch a Falling

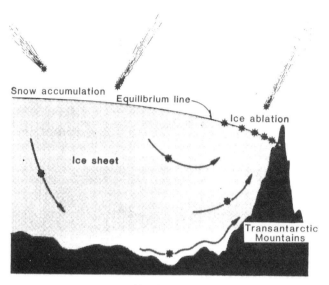

FIGURE 10.10 Profile of the ice sheet showing an idealized model of a meteorite trap. Vertical exaggeration about ×50.

Star: Meteorites and Old Ice." Ian wrote the major part of the paper. As I recall it, I contributed an estimate of the rate of fall of meteorites to the surface and helped with some of the writing.

AN IDEALIZED METEORITE TRAP

We tried to describe an idealized set of conditions consisting of a constant rate of infall of meteorite material over time and a steady-state glacial flow. We postulated an ice flow direction perpendicular to a barrier that extended above the ice surface.

Our model, shown in Figure 10.10, had exposed ice and net ablation occurring at and near the barrier and a snow surface and net snow accumulation over the surface farther from the barrier. There was an equilibrium line where one zone met the other zone, at which neither net ablation nor net accumulation was occurring.

We pointed out that meteorites would fall with equal frequency onto the zone of ice ablation and the zone of snow accumulation, but while the meteorites falling onto the ablation zone would be stranded immediately, meteorites falling into the snow accumulation zone would have a more complicated history. The latter would be buried by accumulating snow. As the snow overburden increased it recrystallized to ice. The included meteorites would then be part of the moving ice sheet traveling toward the zone of net ablation, where they would be uncovered and left stranded on the ice surface, rejoining those fragments that had fallen at the same time, but directly onto the ice surface.

While the Whillans and Cassidy model for meteorite traps was originally set up as a straw man to invite debate, the ideas on which it is based have not successfully been challenged. Everyone agrees, however, that it is overly simple. It was designed originally with the Allan Hills Main Icefield in mind. Ludolf Schultz, of the Max Planck-Institut für Chemie in Mainz, Germany, has pointed out that the flow at Allan Hills is more parallel to, than perpendicular to, the barrier, and the material entering the stranding surface is somewhat incidental to the main direction of flow. At another stranding surface, the ice that forms the western side of the Lewis Cliff Ice Tongue is flowing off the ice plateau of East Antarctica. This flow is diverted from its path by an inert body of ice, the Walcott Névé. Thus, the ice that carries meteorites off the plateau and into the Lewis Cliff Ice Tongue must follow a tortuous route, indeed. Other meteorite traps do not even have physical barriers that extend above the ice surface. So there may be no ideal meteorite trap in the sense that we have proposed it, but hopefully the principles of transport and concentration are the same and can still provide a basis for understanding stranding surfaces and their relation to the ice sheet.

STRATIGRAPHIC SEQUENCES

Obviously, the farther inland a meteorite has fallen, the longer will be its journey to the stranding surface. Its trajectory within the ice will also be deeper, by virtue of the surface above having received more snow accumulation along the way. If one drills upstream of a stranding surface, the drill bit will encounter older and older ice as it

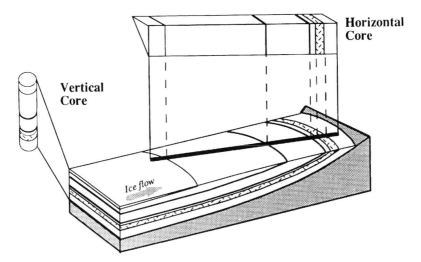

FIGURE 10.11 Schematic diagram showing a horizontal ice core obtained across a meteorite stranding surface compared to a vertical ice core taken inland. Stratigraphic intervals are proportional in both samples.

bites deeper. This deeper ice is proceeding from sources farther and farther inland. Ice of a given age will be characteristic of the same depth below the surface anywhere along the flowline of the ice, *except at and just before the barrier,* where it will be deflected upward (see Figure 10.10). Because of the normal stratigraphic sequence in the ice (i.e., the youngest at the top and the older nearer the bottom), when ice is deflected upward at the barrier, the older ice will emerge nearer the barrier and the younger ice farther away. This should tend to increase the concentration of meteorites with older terrestrial ages closer to the barrier. Another consequence described by this model is that there should be a continuous ice-stratigraphic sequence exposed for sampling across a meteorite stranding surface (Figure 10.11).

Serial sampling of this surface could be carried out using a chain saw. This would be much less expensive than core drilling farther inland. It would have the further advantage that ice at specific levels in a stratigraphic sequence could be sampled in any desired quantity. A possible disadvantage is that diffusion of volatiles might have occurred in the stranding-surface samples due to the fact that the surface ice tends to have a higher fracture density. Ice chemistry and particulate inclusions, however, ought to be undisturbed.

Even though the model postulates an absolute barrier to ice flow, no barrier is infinitely long, so there would be some thinning of the oldest layers at the stranding surface by lateral flow toward the edges of the barrier. This was referred to in the paper by Whillans and Cassidy as "compression," which may not have conveyed the meaning that was intended. Ian's practical analogy was squeezing the contents of a tube of toothpaste against a wall.

DETERMINING ABSOLUTE AGES IN A HORIZONTAL ICE CORE

If a stratigraphic sequence is to be anything more than a relative record, there must be some way to determine the absolute ages of different levels within it. With a vertical ice core this is done by counting yearly layers downward from the surface. The surface, by definition, has zero age. This may not work in a horizontal ice core, because yearly layering may not be detectable, and the level of zero age may not be certain. Zero age in an ideal situation would be found at the equilibrium line between net ablation and net accumulation, but this may be a zone of uncertain width because the equilibrium line undoubtedly oscillates back and forth over the surface as the weather fluctuates.

On rare occasions we have found meteorites embedded in ice just before they become exposed by the ablation process. The terrestrial ages of these meteorites can be determined and the sites where these meteorites are found can become reference points of known age in the stratigraphic column represented by a horizontal ice core. These dates will be useful where they are known, but they would be very few in number.

More useful certainly is the existence of dust bands in the ice. Often these dust bands consist of volcanic glass shards that originally were deposited on a horizontal snow surface over a wide area as a result of a distant volcanic eruption. These bands represent an instant in time whose age may be determined by measurement of their potassium/argon (^{40}K/^{40}Ar) ratio. Glass, having been liquid at eruption, would have retained no argon gas, so any argon present must have accumulated from decay of potassium since the eruption. Dust bands are common enough in ice-ablation zones so that their utility as time markers should be investigated. In fairness, it has to be admitted that these methods of age determination have rather large error limits, so absolute ages in a horizontal ice core would not have the accuracy of those in a vertical core, where seasonal layers can be counted. Figure 10.12 shows a dust band in the Lewis Cliff Ice Tongue.

During the 1994–95 field season, Nelia Dunbar and some colleagues from the New Mexico Institute of Mining and Technology at Socorro, New Mexico, mapped dust bands on the Allan Hills Main Icefield. Their aim was to determine a relative stratigraphic sequence and then to measure ages of the volcanic shards making up the dust. This would give them an absolute measure of the stratigraphic sequence.

I complained earlier that glaciologists seemed to be less than fascinated by meteorite stranding surfaces, but we may now be seeing the beginnings of such interest. Nelia's work has already had the serendipitous result that they have found one of the dust bands to be composed principally of meteorite ablation spherules and meteoritic dust that apparently resulted from the overhead passage of a single H-type chondrite. The terrestrial age of this event has now been measured as 2.3–2.7 million years. In terms of terrestrial age, the oldest meteorite found so far on this stranding surface has a measured age of around 2 million years. This remarkable discovery by glaciologists has greatly strengthened the idea that ancient ice exists at meteorite stranding surfaces, and is easily accessible for sampling.

TERRESTRIAL AGES OF METEORITES AND THEIR STRANDING SURFACES

An inference that could be made is that the oldest meteorites emerging near the barrier would have terrestrial ages directly proportional to their path length between the stranding surface and the point where

they fell. Technically, since finding a meteorite in the act of emerging from the ice is very rare, the terrestrial age of a meteorite on the ice should be greater than the travel time because the meteorite has been lying on the ice for some additional time after reaching the stranding surface.

Kunihiko Nishiizumi, currently at the University of California Space Sciences Laboratory in Berkeley, California, has measured the terrestrial ages of many antarctic meteorites and has found that the terrestrially oldest meteorites at the Allan Hills Main Icefield are all found close to that part of the barrier that rises above the ice surface. Younger ones are mixed with them, but these probably surfaced farther out on the stranding surface and were then transported as windblown debris, or arrived by direct fall onto that part of the stranding surface. Tabulations of meteorite terrestrial ages also indicate that the Allan Hills Main Icefield and the Lewis Cliff Ice Tongue have older meteorites on them than the Yamato Mountains stranding surface and the other known stranding surfaces. At first glance, this may suggest that stranding surfaces have been collecting meteorites for different total lengths of time. I favor an explanation that involves longer trajectories through the ice, as discussed later. As we shall see, however, all known stranding surfaces are much younger than the ice sheet.

The ice sheet of East Antarctica is believed to have been in existence for the last 20 million years. In general, the stranding surface accumulations that we have already discovered contain meteorites most of whose terrestrial ages range from about 10 000 to 300 000 years. The fact that we find accumulations of meteorites that are commonly no older than about 300 000 years is probably a good reason to believe

FIGURE 10.12 A block of ice has been removed with a chain saw, revealing a dust band dipping into the ice. Arrows point to the dust band exposed in the vertical wall of the cut. The dust band consists of volcanic shards. It was deposited horizontally over a large area, and represents a stratigraphic marker, or time horizon. Notice that this dust band dips into the ice, as would be expected in a meteorite stranding surface such as that of Figure 10.11.

that the ice sheet has fluctuated in major ways over the time it has been in existence.

In 1983 I had a paper published with the title, "The remarkably low surface density of meteorites at Allan Hills and implications in this for climate change" (*Proceedings of the Fourth International Symposium on Antarctic Earth Science*). Until then, I had always emphasized the remarkably *high* surface density of meteorites at the Main Allan Hills stranding surface. Both of these concepts are sound: antarctic meteorite concentrations are dazzlingly high compared to concentrations we had found elsewhere in the world, but remarkably low if stranding surfaces had been in existence as permanent features of the ice sheet, i.e., for 20 million years. Using other workers' estimates of about 6 falls/km^2/Myr and making the assumptions that each fall produces an average of 10 fragments and an equal number of fragments are carried into the ablation zone by moving ice, it was easy to calculate that the 75 km^2 of exposed ice at the Main Allan Hills Icefield should accumulate 9000 meteorite specimens in a million years, and 180 000 in 20 million years.

We eventually found about 1200 specimens on the Allan Hills Main Icefield. I interpret this to mean that even if all the assumptions used to estimate the accumulation rate were seriously wrong, the Allan Hills Main Icefield still cannot be as old as the ice sheet. Furthermore, since I consider the assumptions to be pretty reasonable, I would guess that this stranding surface has been operating for no more than about 100 000 years. So it seems certain that Antarctic meteorite stranding surfaces are transitory features, compared to the lifetime of the antarctic ice sheet, probably collecting meteorites for a while just before and during a climatic cold period, and flushing them out during a climate warming stage. At first glance, this idea seems to be incompatible with the terrestrial ages of meteorites found on the stranding surfaces (see Figure 10.13).

While the terrestrial ages of meteorites found on stranding surfaces typically range between 10 000 and 300 000 years, decreasing numbers have ages since fall up to about 2 million years. How to explain the fact that a meteorite that fell 2 million years ago can be

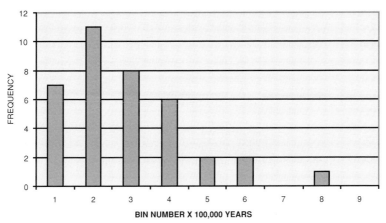

FIGURE 10.13 Terrestrial ages of 37 meteorites from the Allan Hills Main Icefield.

found on a stranding surface that has been collecting meteorites for only 100 000 years? One must assume that this meteorite has spent as much as 1 900 000 years in transit to the stranding surface, traveling rather slowly. In 1 900 000 years, a meteorite can get from the center of the continent to its edge by moving in ice traveling at a 1.2 m/year average speed. This indicates that stranding surfaces collect much more than only the meteorites that fall directly onto them. It also validates the "meteorite trap" model, in which the concentration on the stranding surface is only the end result of a long process that starts with the fall of the meteorite into snow at some, possibly quite distant, upstream point.

Figure 10.13 illustrates the terrestrial age distribution of some meteorites found on the Allan Hills Main Icefield. This tabulation of terrestrial ages has two interesting characteristics. One is that recently fallen meteorites (i.e., those that fell during the last 100 000 years) are less common on the stranding surface than those that fell during the preceding 200 000 years. The other is that, with the exception just mentioned, as their terrestrial ages increase their frequency of occurrence on the stranding surface decreases exponentially.

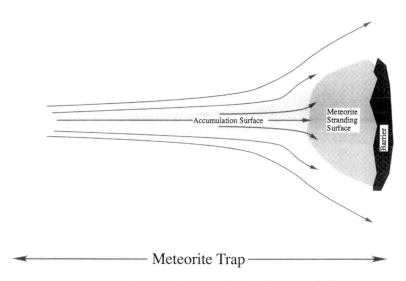

FIGURE 10.14 Schematic diagram of simple flow around a barrier.

The shape of this age distribution curve seems to be characteristic of any stranding surface. The exponential falloff in age has been explained as a weathering phenomenon: the older meteorites have been destroyed by weathering until there are only a few left. This may be a sound concept, but an identical age distribution could be explained as easily by a consideration of the probable shape of the meteorite trap, as sketched in Figure 10.14.

While Figure 10.10 showed an idealized cross-section through a meteorite trap, Figure 10.14 shows an idealized version of that same trap in plan view. Notice that the flowlines that will carry meteorites into the stranding surface narrow in the upstream direction, and the meteorite trap is substantially longer than it is wide. At its extreme upper end, where those meteorites fall that will have the greatest terrestrial age when they reach the stranding surface, the collecting area is at its narrowest. This dictates that a stranding surface will have meteorites with extreme terrestrial ages on it, but only a very few. More are in the pipeline, of course, but that pipeline is very narrow, compared to the zones that will deliver meteorites with younger terrestrial ages, so the specimens with the greatest terrestrial ages will always make up a small proportion of the total number. This model also suggests that when a stranding surface is reestablished at the end of a climate warming stage, meteorites with an ancient time of fall will immediately appear, but typically in very low numbers. The maximum ages will be a function of path length in the ice.

Kunihiko Nishiizumi has pointed out that different stranding surfaces can have meteorites of different average terrestrial age. Allan Hills Main Icefield meteorites, for example, average out to around 250 thousand years and include one sample as old as 2 million years. Yamato meteorites, apparently much younger, average to 75 thousand years and the stranding surface has no meteorites identified as older than 175 thousand years (Nishiizumi, 1990). The probable explanation of this curious fact is not that one stranding surface is younger than another, but that one meteorite trap might extend further into the upstream accumulation zone than another. In the latter case, the trap with the longer path length within the accumulation zone could produce meteorites in the ablation zone that already are older than newly exposed meteorites in the trap with the shorter path length.

The other interesting feature of the age distribution curve is that the younger falls are underrepresented in the frequency spectrum. Remember, however, that the more recent falls may be still enroute to the stranding surface, and will be older by the time they arrive. The specimens represented in the youngest bin, therefore, may have a larger percentage of falls that arrived by direct fall onto the stranding surface.

CONCLUSION

Not being a glaciologist, it would be presumptuous of me to try to infer much from our observations on meteorite stranding surfaces. But I do insist on continuing to think. I continue to encourage glaciologists to interest themselves in the glaciological implications of meteorite stranding surfaces. I continue to miss Ian Whillans' informed collaboration on these questions.

11   The future: what is, is, but what could be, might not

In our model of things, the universe is mainly nothing: an infinite empty space, populated here and there with density nodes of all sizes, from the immeasurably minute, relative to us, to the unimaginably gigantic, relative to us. We seem to be driven to study the universe, and, since the study of nothingness so far has been without profit (except, possibly, to philosophers), we study the density nodes. One aspect of this is the study of meteorites. To study meteorites we must collect examples. To collect examples we have gone to Antarctica.

We have made a good beginning. As of this date (Spring, 2002), there are around 30 000 antarctic meteorite specimens in the combined collections of the US, Japan and Europe. A large fraction of them have not yet been characterized; this is why the *Catalogue of Meteorites* (5th edition) lists only 17 808. If my guess is correct that the antarctic collection averages 10 specimens per fall, 30 000 specimens represents 3000 falls. This compares to around 1000 observed falls in museum collections from the rest of the world. If meteorite finds (3700) are added in, the world's museums have 4700.

Counting the team that has recently returned from the field (austral summer, 2001–02), 25 field expeditions have been mounted by the ANSMET project alone. Since 1977, laboratory space and personnel at the Johnson Space Center have been dedicated to initial processing of the recovered materials and distribution to interested scientists around the world. During the same time, personnel at the US National Museum have been engaged in initial petrological descriptions of recovered material and long-term archiving of material for which current demand has been satisfied. Until recently, virtually all of the initial petrographic descriptions of ANSMET specimens have been carried out by Dr. Brian Mason, at the US National Museum. This alone has been a monumental task!

If we engage in some activity that is worthwhile enough to justify such a significant expenditure of time, talent and resources, we should learn from it. We should also think about directions in which meteorite collecting might proceed in the future. Hence, this chapter.

ANTARCTIC METEORITE FIELD COLLECTION PROGRAMS

With most of the groups listed in Table 9.2 still so rare in our collections that each new specimen may tell us something we did not know about its parent body, we have an obligation to continue. "Rare" is still rare. I am sure there are meteorite groups so exceedingly rare that no example specimens have yet been found; so we also have that possibility to anticipate as we continue to harvest existing meteorite concentrations.

A nice byproduct of this will be that the statistical analysis of some of the more common groups and their metamorphic subgroups will be improved by the addition of numerous new specimens. From time to time, in conversation with colleagues, I have made a tongue-in-cheek suggestion that we continue collecting until the mass–frequency distribution curves of even the rarest groups are unimodal. This will, of course, never be achieved. But it seems worthwhile to continue to move in that general direction.

The potential of known stranding surfaces in Antarctica has not been exhausted – there are meteorite stranding surfaces where our systematic collecting programs have not been completed. Other potential sites are known, but have never been visited, and some of these must contain concentrations of meteorites. It would be tantamount to frittering away a natural resource if these were never recovered.

EXISTING ANTARCTIC METEORITE COLLECTIONS

I incline to the view that the system we have set up to deal with the ANSMET collections is quite good. All collected specimens are characterized as to mass, type, subtype, group and metamorphic sub-group. Research material is made available on a worldwide basis to as many qualified scientists as possible, under a formal program for distribution. As-yet unused material is eventually archived in a controlled environment by a responsible agency, pending any future requests for research specimens. The success of this program depends on a judicious mix of funding, facilities and talented people by the National Science Foundation, NASA and the Smithsonian Institution (US National Museum).

The cost of this special treatment of research material is at least equal to the cost of the field expeditions that recover the specimens. The governmental agencies that fund these endeavors deserve high praise for their willingness to commit the necessary resources to ensure that all possible advantage is gained from the existing collection, and that the collection continues to grow.

TOURISTS COLLECTING METEORITES

Those in Japan, Europe and the United States who are charged with operating the field and curation programs for antarctic meteorites should have some concern about recent threats to the integrity of future collections. These threats involve collection of meteorites by tourists visiting Antarctica, who have gone there for that specific purpose. The locations of meteorite stranding surfaces are known, and if one has enough money he or she can mount a private expedition to collect at these sites. This is being done. It is not likely that the exacting field curation procedures that have been established for collecting antarctic meteorites will be adhered to by these individuals, and mapping the locations of all finds will probably not be done. Even less likely is that post-collection procedures will measure up to those carried out on the existing major collections.

While it is easy to criticize and to view with alarm, it is always difficult to conceive of constructive ways of dealing with a bad situation. I have a sense of empathy with anyone who wants to visit Antarctica, as well as with anyone who wants to hold a meteorite in his or her hand. Maybe it is time to make the antarctic experience available to more of these persons by including them in field teams. Maybe it is also time that the various national programs loosen their hold on pieces of some of the more common meteorites by presenting them to team members who have given their time, in exchange for the pain of extreme cold, in order to be part of an officially recognized meteorite collecting expedition. Such a gesture might entice some of the more ethical private collectors of antarctic meteorites. Aside from these suggestions I can think of no other *non-restrictive, non-legalistic* solution to the problem.

RESEARCH ON EXISTING COLLECTIONS

- Of all the areas in which research is carried out on antarctic meteorites, studies of their cosmic-ray exposure histories and determination of their terrestrial ages seem to proceed most slowly. This is partly because there is only a limited number of scientists with the temperament and the ability to carry out these exacting procedures, and partly because the procedures themselves are so time-consuming. Yet the results are of great interest. Cosmic ray exposure ages measure the length of time an object has been traveling in space as a small body. The extent of decay of cosmic-ray produced radioactive nuclides measures the length of time since the body fell to Earth. Both of these ages give us insights into the history of a meteorite. Additionally, the terrestrial age of a meteorite gives us information about the stranding surface on which it was found and on the history of the ice sheet itself. I expect that these types of age determinations for specimens in the antarctic collections will continue to be made far into the future, and we will be able to draw ever-firmer conclusions about individual stranding surfaces as a result. But it requires patience to wait for the data to appear.

- In Chapter 9, I reported reasons to believe that the flux of HED meteorites at the earth had recently increased. The mechanism I suggested to explain such an observation involves collision of one of the three known earth-approaching Vesta-like asteroids with a body in the asteroid belt. Such a collision could have injected a cloud of debris into earth-crossing orbits. It would be interesting to see if carbon-14 measurements confirm an excess of very short terrestrial ages for this group of achondrites.

- Hopefully, the complete characterization of the Japanese collection will be carried out. This could add very valuable data to the type of statistical treatment I attempted in Chapter 9, even if no more antarctic meteorites were ever collected. Another useful move in the same direction would be the further characterization of those carbonaceous chondrites in the JARE and ANSMET collections that have been designated only as C2.

MICROMETEORITES AND COSMIC DUST

I will use the following terminology. A *micrometeorite* is a tiny meteorite larger than about 50 μm but smaller than about 1–2 mm (these are arbitrary limits). It may have a fusion crust but also may be completely unmelted. *Cosmic dust particles* are extraterrestrial particles smaller than about 50 μm. *Meteoritic spherules* or *ablation spherules* are glassy droplets brushed off meteorites during their fiery plunge through the atmosphere. Possibly some of these are also tiny individuals that were completely melted during deceleration.

Paul Siple, the Boy Scout who accompanied Admiral Richard Byrd's first expedition to Antarctica and returned for his second expedition, had himself lowered on a rope into a cavity in the ice at one of Byrd's Little America base camps. The water supply for Little America had been obtained by melting ice beneath the surface, and this hole in the ice represented the amount of water that had been melted and withdrawn during the entire field season. As melting proceeded, particulate matter that had been embedded in the ice had settled to the floor of the chamber, and Siple collected a jar of these grains. Because the melted ice was below the modern surface, one would expect to find only natural products in the ice – principally volcanic debris from eruptions within a few hundred kilometers of the site, finely ground rock particles picked up from the bed of the glacier, and extra-terrestrial material. There is no record that he collected any rocks, so we can assume he found no possible meteorites. Siple's collection of grains cannot now be located, but it seems a reasonable assumption that a significant fraction of it consisted of micrometeorites, ablation spherules and assorted extraterrestrial dust grains.

In 1986, Michel Maurette, Research Director at the French National Center for Scientific Research, and some colleagues, re-ported finding large quantities of micrometeorites and cosmic dust in meltwater ponds atop the Greenland ice sheet. While examining some aerial photos over Greenland, he had noticed a large number of shallow ponds that seemed to be floored by dark-colored material. With the example of antarctic meteorites lying about on the ice, he had surmised that this dark-colored material might be meteorites. His trip across the ice sheet to the ponds in question must have been an uncomfortable one indeed because he had to ford numerous melt-water streams that typically drain the surface during the summer. When he finally arrived at one of the ponds he was initially quite dis-appointed because the dark-colored "sediments" he had seen on the photos turned out to be algal growth covering the bottom of the pond. Ripping some of it up, however, he found it was rooted in fine sedi-ments, which came up with the algae. The fine sediments turned out to contain a significant fraction of extraterrestrial grains. Apparently the algae were able to thrive just as well on a mixture of windblown volcanic dust and extraterrestrial particles as on any wet, completely terrestrial medium.

In 1987, Maurette and some colleagues processed some 100 metric tons of antarctic ice near the French antarctic research station Dumont d'Urville, expending around 5000 liters of fuel to generate steam which was then injected below the ice surface. They separated about 5000 meteoritic spherules and 5000 unmelted and partly melted micrometeorites by filtering the meltwater. This calcu-lates out to an average of two particles per liter of fuel expended, or one particle per 10 kg of antarctic ice.

During the same field season, Christian Koeberl, of the Geochemistry Institute, Vienna University, Austria and some colleagues collected glacial till samples and successfully separated numerous glassy spherules. During that field season I remember crawling partway up the face of the first ice core moraine at the north end of the Lewis Cliff Ice Tongue, along with a couple of other members of the field team, and sweeping up piles of fine sediment, with which we filled a polyethylene jar. The collection was examined by Ralph Harvey and Michel Maurette. A significant fraction of this material turned out to be extraterrestrial. Apparently the scattering of rocks on the sloping surface created enough turbulence in the wind sweeping down the ice tongue so that tiny grains ablating out of the ice tongue were dropped and left behind as concentrations on the surface.

In 1995, Susan Taylor and some colleagues from the Cold Regions Research and Engineering Laboratory, Hanover, New Hampshire and Ralph Harvey from Case Western Reserve University, Cleveland, Ohio collected fine particles from the meltwater well at South Pole Station. From the vertical dimensions of the cavity it is known that the sampled strata range between 1100 and 1500 AD. Because the Geographic South Pole is so remote from sources of terrestrial contamination, not much volcanic material was expected. Most of the collected material, however, was artificial contamination such as flakes of rust and welding beads from repairs to the heating mechanism. Even so, a vacuum-type collector lowered into the cavity collected about 50 g of tiny extraterrestrial particles. The majority of this collection has not yet been described. Obviously, collecting micrometeorites and other tiny extraterrestrial particles in Antarctica will be a fruitful pursuit for the future.

DOES ROBOTIC COLLECTION OF METEORITES MAKE SENSE?

My last field season with the ANSMET program was the 1993–94 austral summer. In 1995, I was asked to consult with a group at the Field Robotics Center, Carnegie-Mellon University (CMU), who were building a robot to search for meteorites in Antarctica. This was certainly a departure from anything I might have imagined doing! For two years, I walked the five blocks between my office at the University of Pittsburgh and the CMU campus once a week, to meet with this group, which consisted of several faculty members and senior researchers, plus a gaggle of students whose numbers and membership in the group varied somewhat from semester to semester. I discovered that none of the students had ever been to Antarctica and no one in the entire group had ever found a meteorite. The latter is not surprising – how many of us have? This group was characterized by high IQs and pleasant personalities, which I found very stimulating. They radiated optimism in a sort of offhand way like a millionaire giving away dollars that he knows he will never miss. The experience was an enjoyable one and, among the technical terms and acronyms that I could not follow, I was able to insert some observations about the antarctic environment and identifying meteorites on the ice that helped to establish the parameters of what they wanted to do. This was what they wanted me for, and I was able to feel useful.

In 1997 we decided to take some of the components to the ice for testing and in 1998 we took the assembled robot, named Nomad, for a full-dress field rehearsal. The part of Antarctica that we could access at short notice was Patriot Hills, in West Antarctica. Unfortunately, this site does not lie in a part of Antarctica that seems to have meteorite concentrations, so we took along some meteorites to put out on the ice in front of Nomad, to see how smart he was (it's a male). The major question, however, was how he would adapt to the extreme conditions. Nomad had previously been used in the Atacama Desert, in Chile, but could he function at all in extreme cold and blizzard conditions, much less pick out a meteorite from among a group of rocks?

So much of what we read about robots and their capabilities is hyperbole that we tend to start believing it. I recall a TV interview given by Dan Goldin, the NASA Administrator, after launch of the Pathfinder mission to Mars in 1996. He mentioned that Pathfinder, after landing on Mars, was to disgorge a small robot named Sojourner that would be able to move around independently on the surface and collect information. He said, *"It will be like having a 22 lb. geologist on Mars."* He seemed quite pleased with the analogy. I could not decide whether he was intentionally engaging in hyperbole for the sake of publicity, or he really believed what he had said. If the latter, he was demonstrating a woeful lack of understanding of the difference between what a geologist does and what a robot can do. In searching for a similar analogy, I thought of proposing that we assign a desktop computer to run NASA; then we could say that we had a 22 lb. (c. 10 kg) administrator running the agency. But I never got next to Dan to whisper it in his ear and now, of course, he is gone.

As delivered in Antarctica, Nomad's principal tool for distinguishing meteorites from terrestrial rocks was a reflected-light spectrometer that operated only in the visible range of wavelengths. This was a first step. No one pretended to expect infallibility from this instrument. We were able to send Nomad away for 1 or 2 km, where, obedient to remote commands, he took spectra of individuals among two discrete sets of rocks. Each set had some rock types common to both sets and some that were unique to only one set. Examining the results later, in a warm room, we realized that Nomad had told us, "These two sets are somehow different." Great triumphs are often built upon tiny increments such as this.

A graduate student named Liam Pedersen was Nomad's spectrometer person. He spent most of the following year storing rock and meteorite spectra in Nomad's memory. Nomad could then take a spectrum and compare it to spectra in his reference library. The problem was that not all spectra of different rocks are distinguishable from each other. Also, the spectrum obtained would be of a weathered surface. Liam found that he could program Nomad to estimate the probability that a given spectrum represented a meteorite, and this was another incremental step forward.

In 1999, the Division of Polar Programs agreed to take Nomad to Elephant Moraine where John Schutt, from the ANSMET program, joined the CMU team. Most of the team members were already cold-hardened from the previous two field seasons and John is the best meteorite-spotter there is, so it was a good combination. John quickly found several meteorites and placed them among some terrestrial rocks scattered about near the edge of a moraine. He then gave Nomad a rectangular area to search, knowing that it contained three meteorites. Nomad searched the area systematically and successfully identified the meteorites as probable meteorites, with different degrees of probability. It made no false identifications.

This is as far as they got. The money to do all this was not coming from a bottomless pit, and they had reached the bottom. As it stands now, Nomad is slow, it requires maintenance, it cannot unequivocally identify a meteorite and it cannot operate autonomously. An honest comparison between a six-person field team searching for meteorites visually and a six-person field team operating Nomad while Nomad searches for meteorites would reflect very badly on the effectiveness of the latter. In the early days of the space program we had a series of failures and it became a little comical when a NASA spokesperson would end every summary of the disaster with the words, "... but we learned a lot." Maybe in some cases all they learned was that they had to try again, perhaps with some new ideas.

In thinking about the experience, I have concluded that success should not be measured by whether or not a robot would be better than a field team in searching for antarctic meteorites. Antarctica, after all, is a great proving ground (for humans, as well as machines), and we can anticipate a need for all kinds of robotic equipment, operating autonomously as well as in the guise of a field assistant in all kinds of environments. Maybe the better approach is to start by designing autonomous devices that perform simple functions in the antarctic environment. Following are some ideas.

- One could start with a self-powered specimen chest. As a team member finds meteorites and collects them a nearby specimen

chest could trundle over and stand by, ready to receive the
packaged specimen. It could register the field number and determine the geographic coordinates of the find by means of its
self-contained GPS instrument, while the team member continues his or her search.

- A small, semi-autonomous radar set. A meteorite stranding surface can be as much as 50% covered by snow dunes migrating
slowly across the ice (see, for example, Figure 10.1, of the Allan
Hills Main Icefield). These may be as much as 2 to 30 cm thick,
from the edge to the center. On a stranding surface that is 50%
covered by snow dunes, it seems reasonable to suppose that 50%
of the meteorites will not be found, except by multiple visits
during which the dunes migrate off previously covered areas.
A self-powered radar set with the antenna looking vertically
downward can be assigned an individual snow dune to probe,
marking the locations of any rocks it might detect under the
snow. Once finished, it could be assigned another snow dune,
or locate one on its own.

- A somewhat more complicated field assistant arrangement
would be a tandem radio echo-sounding device for measuring
ice thickness and bottom contours of meteorite stranding surfaces. This was done at the Lewis Cliff Ice Tongue by traditional
methods, and helped to explain why meteorites were preferentially concentrated on the western side of the northward-
flowing ice tongue. The proposed system would require two
semi-autonomous robots operating in tandem. Each would drag
an antenna wire; one to transmit a radar pulse into the ice and
the other to receive its reflection from the bottom. Locations of
measurements would be recorded by GPS units in each of the
vehicles. This field assistant team could operate independently
of human team members searching for meteorites on the same
stranding surface.

- A very useful advanced device is a roving multispectral scanner
that would analyze reflected light in the visible, near infra-red

and far infra-red regions of the spectrum. Mineral identification becomes easier in proportion to the number of absorption lines that can be detected. Nomad collected spectra only in the visual range, and had significant successes even in spite of this limitation.

- An autonomous power station. Once assigned to a stranding surface where a human team expects to spend two to six weeks, the robots described above would not have to return to camp at all, if there were a nearby power source to recharge their batteries whenever needed. Such a power source would be a structure supporting solar panels or an array of windmills, or both. The power station itself could be capable of locomotion. It could follow in a general sense the locus of activity of operating robots, so they would not have to travel very far for a recharge. It could position itself more advantageously to follow the sun and also to avoid being mired in drifting snow during storms.

Success with the relatively simple single-function machines I suggest above could lead to integration of functions in later models, so that development of more sophisticated robots could proceed in an evolutionary manner that would yield increasing benefits to the meteorite collecting teams. This could result, finally, in a much-improved Nomad (Nomad II). The improved model might be able to tell the difference between terrestrial rocks and meteorites by screening the data according to spectrum, surface texture, magnetic properties, and so on. It might be able to collect the meteorites it found; then summon a specimen box to its side. Nomad II could be set to carry out tasks such as the intellectually unrewarding (read: deadly dull) process of searching moraines for the few meteorites that might be hiding among thousands of rocks.

The kind of robotics program I suggest would also have another agenda item – along with teaching machines to deal with the nasty antarctic environment, pioneering a line of complex robots to be used in *really* nasty environments.

MARS

The martian ice caps may be older than the antarctic ice sheet. They seem to be composed of water ice and carbon dioxide ice, and they may have been receiving meteorites and meteoritic dust for a very long time. The only other materials one would expect to be present there are martian windblown dust and impactites from martian impact craters. One cannot predict what the ratio of meteorites to impactites would be, but the meteorites might greatly outnumber the impactites. If that were so, a robotic expedition over a martian ice cap need only collect every rock it could find in order to amass a sizable collection of meteorites.

Mars is much closer to the asteroid belt than is Earth. Has it been receiving a higher flux of meteorites over time? Having collected for a longer time, does it have a higher percentage of the rarer meteorite groups? These are interesting possibilities, and there are no current answers to these questions, so a robot that can identify a meteorite on the martian surface could be worth having there.

LAGRANGIAN POSITIONS

As described long ago by the mathematician Lagrange, the orbit of a planet traveling about the sun has two points that are equidistant from both the sun and the planet (lagrangian points). One lagrangian point precedes the planet in its orbit and the other trails. A body located in one of these points is located at the third vertex of an equilateral triangle whose other vertices are the sun and the planet. Because of the opposing forces acting upon it (gravitational attraction of the planet, gravitational attraction of the sun and centrifugal force), a body occupying one of these points (zones, really) tends to remain there.

Nothing has been observed telescopically at the lagrangian points of the earth, but small undiscovered meteoroids may have been trapped there, so perhaps these positions could be visited by a robotic collector using the International Space Station as a base of operations.

A family of asteroids has been identified (the Trojan asteroids) that occupy the lagrangian positions of Jupiter. These must be bodies that were on a path to be thrown out of the solar system or were asteroids that had formed farther out and were being perturbed inward when they happened to pass into the gravity well of one of Jupiter's lagrangian points, without the energy to climb out. It would be of much interest to see if these asteroids are the same, or different from, the parent bodies that produced the meteorites we are familiar with.

Certainly, we should be thinking of constructing robots to land on and sample the surface materials of asteroids, both in Jupiter's lagrangian positions and throughout the asteroid belt. An important question that has been so far unanswered is, "As evidenced by the bodies that formed there, did the average composition vary across the asteroid belt?"

There is much to do. Let us get busy.

Appendix A **The US–Japan agreement**

Text of the memorandum that served as an agreement between Takesi Nagata and William Cassidy for cooperation in a field search for meteorites.

TO: DUWAYNE ANDERSON

FROM: DR. T. NAGATA
            DR. W. CASSIDY

DATE: DECEMBER 9, 1976

SUBJECTS: 1. POSSIBLE RECOVERY OF METEORITE SPECIMENS RESULTING FROM JOINT FIELD EXCURSIONS DURING 1976–77 IN THE DRY VALLEY AREA OF ANTARCTICA.
           2. DISPOSITION OF RECOVERED SPECIMENS.
           3. PROVISION FOR EXPANDING THE INITIAL SCOPE OF THE WORK.
           4. RESEARCH PROGRAMS

1. Logistics and base facilities of the USARP program at McMurdo will be used by a joint U.S.–Japan team to search for meteorites in the Dry Valleys and adjacent parts of the surrounding ice cap during the 1976–77 field season.

2. Any meteorite specimens recovered will be distributed in the following way:

   a. Specimens larger than 300 g will be cut in two approximately equal pieces at the Thiel Earth Science Laboratory (in McMurdo). One piece will be utilized by the U.S. group and the other by the Japan group.

b. Specimens 300 g or smaller will be distributed in equal numbers between the groups on an alternate-choice basis. Each group will retain the privilege of later requesting study materials from the other group's collection in connection with existing research programs of their own.

3. As observations from helo pilots and other groups come in, we may find it desirable to visit other field areas. The arrangements described above will apply to any meteorites recovered as a result of such change of plans.

4. Even though specimens will be distributed between our two groups we will remain in contact about our current research programs on them, in order to avoid duplication of effort and in order to plan better how they may be utilized. We feel it would be appropriate to acknowledge the efforts of the joint U.S.–Japan team in any subsequent publication of research results.

Signed: W.A. Cassidy

                                                          Takesi Nagata

Appendix B **ANSMET field participants 1976–1994**

Most of the following records are taken from a list tabulated by John Schutt which he called "List of ANSMET Field Members through the Ages." I have included only those field seasons during which I was Principal Investigator – not because field members in subsequent seasons are not honorable, or do not deserve to be honored, but because any such list in a continuing project will never be complete, and one must stop somewhere. Affiliations are those of each member at the time they went to the field.

1976–77: Edward Olsen, Field Museum of Natural History, Chicago, IL
        Keizo Yanai, National Institute of Polar Research, Tokyo, Japan

1977–78: Billy P. Glass, Dept. of Geology, University of Delaware, Newark, DE
        Minoru Funaki, National Institute of Polar Research, Tokyo, Japan
        Keizo Yanai, National Institute of Polar Research, Tokyo, Japan

1978–79: John Annexstad, NASA/Johnson Space Center, Houston, TX
        Dean Clauter, University of Pittsburgh, Pittsburgh, PA
        Minoru Funaki, National Institute of Polar Research, Tokyo, Japan
        Everett Gibson, NASA/Johnson Space Center, Houston, TX
        Steven Hartmann, Antarctic Support Division, Holmes and Narver, Inc.
        Ursula Marvin, The Smithsonian Astrophysical Observatory, Cambridge, MA
        Fumihiko Nishio, National Institute of Polar Research, Tokyo, Japan

Kazuyuki Shiraishi, National Institute of Polar Research, Tokyo, Japan

1979–80: John Annexstad, NASA/Johnson Space Center, Houston, TX

Lee Benda (crevasse expert, independent contractor)

Fumihiko Nishio, National Institute of Polar Research, Tokyo, Japan

Louis Rancitelli, Batelle Memorial Institute, Columbus, OH

1980–81: John Annexstad, NASA/Johnson Space Center, Houston, TX

Joanne Danielson, The University of Pittsburgh, Pittsburgh, PA

Harry (Hap) McSween, The University of Tennessee, Knoxville, TN

Louis Rancitelli, Batelle Memorial Institute, Columbus, OH

Ludolf Schultz, Max Planck-Institut für Chemie, Mainz, Germany

John Schutt (crevasse expert, independent contractor)

1981–82: Ghislaine Crozaz, Washington University, St. Louis, MO

Robert Fudali, The Smithsonian Institution/USNM, Washington, DC

Ursula Marvin, The Smithsonian Institution/SAO, Cambridge, MA

John Schutt (crevasse expert, independent contractor)

Ian Whillans, The Ohio State University, Columbus, OH

1982–83: John Annexstad, NASA/Johnson Space Center, Houston, TX

Kristine Annexstad, Rice University, Houston, TX

Vagn Buchwald, Danmarks Tekniske HØjskole, Lyngby, Denmark

Richard Crane, United States Navy

Urs Krähenbuhl, Universität Berne, Bern, Switzerland

Tony Meunier, The U. S. Geological Survey, Reston, VA

Louis Rancitelli, Batelle Memorial Institute, Columbus, OH

John Schutt (crevasse expert, independent contractor)

Carl Thompson (crevasse expert, independent contractor)

1983–84: Robert Fudali, The Smithsonian Institution/USNM, Washington, DC

A. C. Hitch, (mechanic, no affiliation)

Kunihiko Nishiizumi, The University of California, San Diego, La Jolla, CA

Paul Pellas, Musée Nacional d'Histoire Naturelle, Paris, France

Ludolf Schultz, Max Planck-Institut für Chemie, Mainz, Germany

John Schutt (crevasse expert, independent contractor)

Paul Sipiera, William Rainey Harper College, Palatine, IL

1984–85: Catherine King-Frazier, James Madison University, Harrisonburg, VA

Scott Sandford, NASA/Ames Research Center, Moffett Field, CA

John Schutt (crevasse expert, independent contractor)

Roberta Score, NASA/Johnson Space Center, Houston, TX

Carl Thompson (crevasse expert, independent contractor)

Robert Walker, Washington University, St. Louis, MO

1985–86: Peter Englert, San Jose State University, San Jose, CA

Ludolf Schultz, Max Planck-Institut für Chemie, Mainz, Germany

John Schutt (crevasse expert, independent contractor)

Twyla Thomas, The Smithsonian Institution/USNM, Washington, DC

Carl Thompson (crevasse expert, independent contractor)

Ernst Zinner, Washington University, St. Louis, MO

Michael Zolensky, NASA/Johnson Space Center, Houston, TX

1986–87: Christian Koeberl, Universität Wien, Vienna, Austria

Louk Lindner, Rijksuniversiteit te Utrecht, Utrecht, The Netherlands

Austin Mardon, Texas A. & M. University, College Station, TX

John Schutt (crevasse expert, independent contractor)

Keizo Yanai, National Institute of Polar Research, Tokyo, Japan

1987–88: Joan Fitzpatrick, U.S. Geological Survey, Denver, CO

Robert Fudali, The Smithsonian Institution/USNM, Washington, DC

Ralph Harvey, The University of Pittsburgh, Pittsburgh, PA
Gary Huss, California Institute of Technology, Pasadena, CA
John Schutt (crevasse expert, independent contractor)
Carl Thompson (crevasse expert, independent contractor)
Faith Vilas, NASA/Johnson Space Center, Houston, TX
Jerry Wagstaff, NASA/Johnson Space Center, Houston, TX
Peter Wasilewski, NASA/Goddard Space Flight Center, Greenbelt, MD

1988–89: David Blewett, University of Pittsburgh, Pittsburgh, PA
Monica Grady, Open University, Milton Keynes, UK
Ralph Harvey, The University of Pittsburgh, Pittsburgh, PA
Randy Korotev, Washington University, St. Louis, MO
Scott Sandford, NASA/Ames Research Center, Moffett Field, CA
John Schutt (crevasse expert, independent contractor)
Roberta Score, NASA/Johnson Space Center, Houston, TX

1989–90: *Field season cancelled.*

1990–91: Mario Burger, Universität Berne, Bern, Switzerland
Ghislaine Crozaz, Washington University, St. Louis, MO
Sue Ivison, ITT Antarctic Services, Inc., Paramus, NJ
John Schutt (crevasse expert, independent contractor)
Suzanne Traub-Metlay, University of Pittsburgh, Pittsburgh, PA
Robert Walker, Washington University, St. Louis, MO
Peter Wasilewski, NASA/Goddard Space Flight Center, Greenbelt, MD

1991–92: Francisco Anguita, Universidad Complutense, Madrid, Spain
Ralph Harvey, The University of Pittsburgh, Pittsburgh, PA
Alexander Krot, UCLA, Los Angeles, CA
Thomas Meisel, Universität Wien, Vienna, Austria
John Schutt (crevasse expert, independent contractor)
Peter Wasilewski, NASA/Goddard Space Flight Center, Greenbelt, MD
Michael Zolensky, NASA/Johnson Space Center, Houston, TX

1992–93: Sue Ivison, Antarctic Support Associates, Englewood, CO

        Jerry Delaney, Rutgers University, New Brunswick, NJ

        Ralph Harvey, University of Tennessee, Knoxville, TN

        Peter Mouginis-Mark, University of Hawaii, Manoa, HI

        John Schutt (crevasse expert, independent contractor)

        Meenakshi Wadhwa, Field Museum of Natural History, Chicago, IL

1993–94: Julius Dasch, NASA Headquarters, Washington, DC

        Keith Echelmeyer, University of Alaska, Fairbanks, AK

        Ralph Harvey, University of Tennessee, Knoxville, TN

        Candace Kohl, The University of California, San Diego, La Jolla, CA

        Sara Russell, The California Institute of Technology, Pasadena, CA

        John Schutt (crevasse expert, independent contractor)

Index of people

Index of Antarctic geographic names

Note: page numbers in bold indicate a main section or full details in text.

Subject index

Note: page numbers in bold indicate a main section in the text or appendix or full details.